T0140870

New Light Sources for Quantum Information Processing

Single Photons from Single Quantum Dots and Cavity-Enhanced Parametric Down-Conversion

DISSERTATION

zur Erlangung des akademischen Grades
doctor rerum naturalium
(Dr. rer. nat.)
im Fach Physik

eingereicht an der
Mathematisch-Naturwissenschaftlichen Fakultät I
Humboldt-Universität zu Berlin

von
Herrn Dipl.-Phys. Matthias Scholz
geboren am 28.05.1979 in Hannover

Präsident der Humboldt-Universität zu Berlin:
Prof. Dr. Dr. h.c. Christoph Markschies

Dekan der Mathematisch-Naturwissenschaftlichen Fakultät I:
Prof. Dr. Lutz-Helmut Schön

Gutachter:

1. Prof. Dr. Oliver Benson
2. Prof. Dr. Thomas Elsässer
3. Prof. Dr. Shigeki Takeuchi

eingereicht am: 22. Januar 2009
Tag der mündlichen Prüfung: 23. März 2009

Bibliografische Information der Deutschen Nationalbibliothek

Die Deutsche Nationalbibliothek verzeichnet diese Publikation in der
Deutschen Nationalbibliografie; detaillierte bibliografische Daten sind
im Internet über http://dnb.d-nb.de abrufbar.

ISBN 978-3-8325-2220-9

Logos Verlag Berlin GmbH
Comeniushof, Gubener Str. 47,
10243 Berlin
Tel.: +49 030 42 85 10 90
Fax: +49 030 42 85 10 92
INTERNET: http://www.logos-verlag.de

Zusammenfassung

Die intensive Forschung auf dem Gebiet der Quanteninformationsverarbeitung spiegelt ihre Bedeutung für zukünftige Anwendungen in den Informationswissenschaften sowie für ein besseres Verständnis fundamentaler Fragestellungen der Quantenmechanik wider. Diese Arbeit beschäftigt sich mit dem Photon als wichtiger Ressource für Aufgaben in der Quanteninformation und mit der Kodierung von Quantenbits auf seine vielfältigen Freiheitsgrade. Seine schwache Wechselwirkung mit der Umgebung macht das Photon zum optimalen Träger, um Information sicher mit Hilfe quantenkryptographischer Protokolle zu übertragen. Zudem erlaubt es die Implementierung von Quantenalgorithmen allein unter Verwendung linearer optischer Komponenten. Skalierbarkeit ist mit einem rein photonischen Ansatz jedoch nicht einfach zu erreichen, da die Erzeugung ununterscheidbarer Einzelphotonen von mehreren Emittern weiter eine schwierige Aufgabe darstellt. Für komplexere Quantennetzwerke wurde daher vorgeschlagen, einzelne Photonen lediglich als Informationsträger zwischen atomaren Ensemblen zu verwenden, die ihrerseits als Speicher- und Prozessierungsknoten agieren. Algorithmen, die nur eine geringe Anzahl von Quantenbits erfordern, können dabei dennoch mithilfe des Schemas der linearen Optik gelöst werden.

In dieser Arbeit wird daher zunächst die Erzeugung und Charakterisierung von Einzelphotonenzuständen diskutiert. Ein optisch gepumpter einzelner Quantenpunkt wird als Einzelphotonenquelle verwendet, um zum ersten Mal die Anwendbarkeit dieser Photonen für die Implementierung eines Quantenalgorithmus mit linearer Optik und Einzelphotoneninterferenz zu zeigen. Fehlerkorrektur macht den interferometrischen Aufbau dabei robust gegen Phasenrauschen. Auf diese erfolgreiche Demonstration eines Modellexperimentes folgt eine Arbeit zur Entwicklung einer Einzelphotonenquelle für praktische Anwendungen. Besonders die Quantenkryptographie, heutzutage die am weitesten fortgeschrittene Quanteninformationstechnologie, benötigt kompakte und preiswerte nichtklassische Lichtquellen. Dazu wird eine auf elektrisch gepumpten Quantenpunkten basierende Einzelphotonenquelle vorgestellt, die unerreichte spektrale Reinheit und nichtklassische Photonenstatistik aufweist.

Die Realisierung von Quantennetzwerken wird in den folgenden Kapiteln verfolgt, die sich mit der Erzeugung schmalbandiger Einzelphotonen befassen, um effizient an atomare Resonanzen zu koppeln. Photonen mit der spektralen Breite atomarer Übergänge werden dabei durch parametrische Frequenzmischung mit Resonatorüberhöhung erzeugt, und ihre Quantenstatistik wird detailliert untersucht. Ein Aufbau zur *time-bin* Kodierung wird vorgestellt, mit dessen Hilfe diesen schmalbandigen Einzelphotonen Quanteninformation aufgeprägt werden kann. Das abschließende Kapitel untersucht als Vorstudie zur Atom-Photon-Wechselwirkung im Einzelphotonenregime die Abbremsung von kohärentem Licht in atomaren Ensemblen.

Die beschriebenen Experimente zeigen eindrucksvolle Eigenschaften, die das Photon zu einem bemerkenswerten physikalischen System für die Quanteninformationsverarbeitung machen.

Summary

The outstanding research efforts in quantum information processing over the past two decades reflect the promise this field of physics provides for practical applications in information science as well as for new approaches towards a better understanding of fundamental questions in quantum mechanics. This thesis focuses on the photon as a principal resource to perform quantum information tasks and on schemes to imprint quantum bits onto its various degrees of freedom. Its weak coupling to the environment makes the photon an ideal carrier to securely transmit information by quantum cryptographic protocols. Moreover, efficient implementations of quantum computing using solely linear optics have been proven. Unfortunately, scalability is not easily achieved by a purely photonic approach since the generation of indistinguishable single photons from multiple emitters remains a difficult task. Thus, proposals for more complex quantum networks suggest an architecture with single photons as information carriers between atomic ensembles that act as storage and processing nodes. Computations including a limited number of qubits, however, may be performed by the linear optics scheme.

The thesis starts with the generation and characterization of single-photon states, using a source based on a single optically pumped quantum dot. The capability of these states to implement a quantum algorithm using linear optics and single-photon interference is experimentally demonstrated for the first time. Error correction makes the interferometric setup robust against phase-noise. After successful realization of this proof-of-principle experiment, attention is drawn to the need of plug-and-play single-photon sources. Especially quantum key distribution, the most advanced quantum information technology to date which has even found its way into commercial devices, requires compact and low-cost non-classical light sources. Therefore, a single-photon source based on electrically pumped quantum dots is presented that exhibits unmatched spectral purity and single-photon statistics.

Results towards the realization of quantum networks are presented in the following chapters, covering the generation of narrow-band single photons which can efficiently couple to atomic resonances. Photons with a spectral width of less than 3 MHz are created by ultra-bright cavity-enhanced spontaneous parametric down-conversion, and their quantum statistics is studied in detail. A setup for time-bin encoding is demonstrated, capable of imprinting quantum information onto these narrow-band single photons. This thesis concludes with slow-light experiments in atomic ensembles as a model system for atom-photon interaction on the single-photon level. The described experiments demonstrate striking features that make the single photon one of the most remarkable physical systems for the field of quantum information.

Dedication

Dedicated to the memory of my mother and my grandmother.

Contents

Chapter 1

Introduction

"Seit 50 Jahren grüble ich darüber nach
was ein Lichtquant sei, und kann es immer
noch nicht sagen. Heute glaubt jeder Lump
er wüsste es – aber er weiß es nicht."
– Albert Einstein

The quantum nature of the harmonic oscillator was introduced by Planck in 1900 in an attempt to account for the characteristics of black body radiation at thermal equilibrium, but he considered quantization a mathematical device rather than attributing physical significance to this assumption. It took another five years before Einstein developed his hypothesis of light quanta or *photons* as a means of explaining the photoelectric effect. He also noticed that Planck's idea could be applied to solve the "ultraviolet catastrophe" in the Rayleigh-Jeans law – a term that was used by Ehrenfest in 1911 for the first time. Originally starting as an academic construct, the photon has become a workhorse to test the foundations of quantum physics against recurring efforts of a purely classical interpretation of nature [1–3], and quantum mechanics advanced to an indispensable tool to drive crucial technological innovations like the transistor [4] or the laser [5].

In the early eighties, fundamental concepts of quantum mechanics were applied to information theory [6]. A new branch evolved within quantum physics, called *quantum information processing* (QIP). Analogously to the classical bit, the *quantum bit* or *qubit* was introduced as a central item to quantify information. It can be interpreted as a two-level system, described by a coherent superposition of two eigenstates. This thesis focuses on the

1

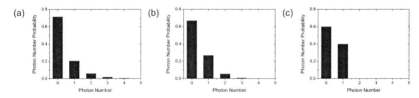

Figure 1.1: Photon number distributions of (a) thermal light, (b) a coherent state as emitted by a laser well above threshold, and (c) a single-photon source, each with a mean photon number $\bar{n} = 0.4$.

single photon as a promising resource for QIP which entered the stage to play an important role as an ideal tool to transmit quantum information over long distances due to its weak coupling to the environment. In 1984, Bennett and Brassard proposed a protocol for secret key distribution [7] that uses the single-particle character of a photon to avoid any possibility of eavesdropping on an encoded message (for a review see [8]).

Additionally, QIP offers the efficient implementation of algorithms which – according to today's state of knowledge – show exponential scaling by classical techniques, like searching unordered databases [9] or factoring [10]. Knill, Laflamme, and Milburn proposed efficient *quantum gates* based solely on the interference of indistinguishable single photons and linear optics [11] which was experimentally demonstrated shortly after [12, 13]. Other proposals extend the role of photons to *flying qubits* as information carriers in scalable *quantum networks* [14–16] between processing nodes consisting of *stationary qubits*, like ions [17, 18], atoms [19], quantum dots (QDs) [20], or Josephson qubits [21, 22].

Linear optics applications in QIP require the reliable deterministic generation of single- or few-photon states. However, due to their bosonic character, photons tend to appear in bunches. Thus, classical light sources provide a broad photon number distribution, as depicted for thermal and laser light in figures 1.1(a) and 1.1(b), respectively. This characteristics hinders the application of classical sources particularly to quantum cryptographic systems since an eavesdropper may gain partial information by a beam splitter attack. Similar obstacles occur for linear optics quantum computation (LOQC) where photonic quantum gates [11], *quantum repeaters* [23], and *quantum teleportation* [24] require the preparation of single- or few-photon states *on demand* in order to obtain reliability and high efficiency. While an ideal non-classical single-photon source emits a sub-Poissonian photon number distribution with exactly one photon at a time, real sources have inevitable losses due to scattering and absorption which lead to typical photon number distributions as shown in figure 1.1(c) [25]. A promising process for single-photon generation is spontaneous emission of a single quantum emitter. Numerous emitters have been used to demonstrate single-photon emission [26]. Single atoms or ions are the most fundamental systems. They have been trapped and coupled to optical resonators to obtain single-mode emission [27, 28] and to increase the collection efficiencies. Other systems capable of single-photon generation are single molecules and nanocrystals [29–31]. However, their drawback is their susceptibility for photobleaching and blinking [32, 33]. Stable alternatives are nitrogen-vacancy defect centers in diamond [34, 35], but they show broad optical spectra together with comparably long lifetimes.

The first part of this thesis will cover single-photon generation from single self-assembled QDs and its application to tasks in QIP. QDs are few-nanometer sized semiconductor structures which resemble features known from atoms, like discrete emission spectrum and electronic structure, and which are therefore often referred to as *"artificial atoms"*. Most experiments with QDs have to be conducted at cryogenic temperature in order to reduce electron-phonon interaction and thermal ionization. High count rates can be obtained due to their short transition lifetimes, and their spectral lines are nearly lifetime-limited. With different material systems, the ultraviolet, visible, and infrared spectrum can be covered. QDs also gain attractiveness by the possibility of electric excitation [36] and the implementation in integrated photonic structures [37].

Most of the described single-photon sources show a wide spectral bandwidth, except those based on atoms and ions. Additionally, all of them emit isotropically, if not implanted into a resonator, so only a fraction of the generated photons is usable in subsequent experiments. Both problems can be avoided by a source based on *spontaneous parametric down-conversion* (SPDC). Inside a nonlinear crystal, a pump photon is split in so-called *signal* and *idler* photons [38–40] which can be correlated in multiple degrees of freedom, like polarization, phase, or frequency. Momentum conservation determines their direction of propagation. Taking advantage of improved fabrication techniques for nonlinear crystals, SPDC has become a standard tool of quantum optics allowing easy generation of entangled photon pairs [41]. Since signal and idler photons are created simultaneously [42], the detection of the first *heralds* the presence of the latter. At low conversion rates and for short coherence times, this strong correlation allows for the application of this technique to heralded single-photon generation. However, coherence times of usually ~ 100 fs translate into THz bandwidth. Aiming for an interface between stationary atomic states of MHz width and flying single photons for the realization of quantum networks, the parametric fluorescence needs to be

3

spectrally filtered, leading to a substantial drop in count rate and inefficient atom-photon interaction.

A promising road around this problem is the resonant enhancement of the parametric fluorescence inside an optical resonator. Thus, part II of this thesis starts with the demonstration of narrow-band single-photon sources based on cavity-enhanced SPDC that keep the overall count rate stable and are even capable to increase it. This resonator configuration is termed *optical parametric oscillator* (OPO). An OPO presents a tunable source of coherent light for spectroscopic applications in the above-threshold regime and allows the generation of non-classical states, exhibiting single-photon statistics far below or squeezing near threshold. OPOs far below threshold have been demonstrated by several groups [43–46]. None of them, however, could prove the single-photon character of a signal or idler field generated by a type-II process. Orthogonal polarizations of signal and idler photons are required to separate both fields behind the source in order to use one of them for heralding while the other is directed towards a subsequent experimental task. For efficient atom-photon interaction, the single-photon state must be long-term stable and needs to consist of a single longitudinal OPO mode. These problems can be tackled by locking the OPO cavity to the SPDC pump beam and by adding a passive filter assembly behind the narrow-band single-photon source.

With a reliable single-photon source of MHz bandwidth in operation, experiments towards efficient coupling between stationary and flying qubits are possible as outlined in figure 1.2. Since atom-photon coupling on the single-photon level is a basic ingredient of *quantum repeater* protocols [23] that aim for long-distance quantum communication, quantum information can be imprinted onto the narrow-band photon state by the concept of *time-bin* encoding [47–50] that surpasses the robustness of polarization-encoded states in optical fibers. For narrow-band photons with wave packet extensions of nearly 100 m, this technique makes great demands on the stability of the applied interferometers, and chapter 7 provides an experimental realization. Smaller computational tasks, which require a limited number of qubits, may be solved by applying LOQC.

Quantum networks, that combine atomic ensembles and single photons, can utilize the joint advantages of these two physical systems. Interfaces between atomic and photonic systems are significant for network synchronization and the storage of quantum information for processing purposes [51–53]. The transfer of photonic states onto atoms can be realized by the coherent process of *electromagnetically induced transparency* (EIT) [54, 55]. Pre-studies of this coherent effect in hot cesium vapor will complete part II of this thesis. Cesium vapor is planned to serve as a quantum relay in this experiment

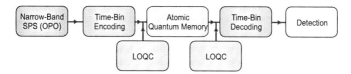

Figure 1.2: Overall scheme for the realization of atom-photon coupling on the single-photon level. Narrow-band single photons are created by a single-photon source (SPS) based on cavity-enhanced parametric down-conversion, and quantum information is robustly imprinted onto them by time-bin encoding in an interferometric setup. Behind an atom-photon interface based on the concept of electromagnetically induced transparency, this information can be retrieved by interferometric techniques.

on the long term, and frequency matching between the narrow-band single-photon source and an atomic transition is inevitable. Beyond the slow group velocities of light pulses using static EIT, light can be brought to a complete halt in order to realize a *quantum memory*. This technique, called dynamic EIT, can theoretically be explained by the concept of *dark-state polaritons* as introduced by Fleischhauer and Lukin [56, 57]. Dynamic EIT can be extended to non-classical states [58], and single-photon statistics is preserved during storage [59]. In 2005, single photons have been stored in atomic ensembles for the first time [60, 61]. Single-photon sources based on an atomic system themselves were used to avoid bandwidth conflicts. The application of cavity-enhanced SPDC, however, means a purely optical approach and allows the combination of different physical systems capable of QIP – a task that will further help to boost the development of larger qubit systems. Beside atom-photon interaction, narrow-band single photons offer fundamental studies of the non-locality of single-photon wave packets due to their comparably long spatial extension.

For future implementations of scalable quantum communication [23] and measurement-based quantum computation [11, 62], it is essential to combine quantum teleportation [63] and quantum memory [16, 64–66] of photonic qubits. After first proof-of-principle experiments tackling these tasks separately [24, 60, 61, 67–69], memory-built-in teleportation of photonic qubits was demonstrated [70] where an unknown polarization state of a single photon was teleported onto a remote atomic qubit that also served as a quantum memory. This teleportation between photonic and atomic qubits with the direct inclusion of a readable quantum memory represents an additional step towards the long-term goal – the realization of efficient and scalable quantum networks.

Part I

Quantum Dots as Single-Photon Sources

Chapter 2

Quantum Computation with Single Photons

"Über Halbleiter sollte man nicht arbeiten,
das ist eine Schweinerei,
wer weiß, ob es überhaupt Halbleiter gibt."
– Wolfgang Pauli

The introductory chapter offered an overview of diverse research efforts in the field of quantum information processing. For larger quantum network architectures, the interplay of flying photonic and stationary atomic qubits could be identified as a promising road towards long-distance quantum communication and computation. For smaller computational tasks, though, the seminal proposal by Knill, Laflamme, and Milburn (KLM) [11] shows an alternative implementation of quantum circuits, based on the interference of single photons in a linear optics setup. This chapter will give an introduction to single-photon statistics, single-photon generation from epitaxially grown semiconductor quantum dots, and their application in a demonstration of the Deutsch-Jozsa algorithm [71, 72], showing principal building blocks of the KLM proposal like single-photon interference and quantum gate operation on the single-photon level. Complex quantum computational setups will also require schemes for error correction; the experimental realization is therefore completed by the demonstration of noise tolerant encoding of qubits in decoherence-free subspaces.

9

2.1 Single-Photon Statistics

Single photons are the physical system of choice to transmit quantum information over long distances and also allow the efficient implementation of small quantum algorithms. A wide variety of systems is capable of generating single-photon states; thus, a universal measure is required to characterize the quantum nature of their photon statistics and to choose the best physical system for a selected task.

Temporal intensity correlation functions are used to provide this measure. The first-order correlation function computes coincidences between field amplitudes, but the probability amplitudes or the two paths of a single photon interfere like classical waves. Therefore, it cannot be considered the proper figure of merit to distinguish between the classical or quantum nature of a light source which has been the subject of a gedankenexperiment by Taylor in 1909 [73]. The second-order correlation function, however, accounts for the required intensity correlations. If a certain mode a of the electric field is correlated with itself, it is often termed *auto-correlation* function and – after normalization – takes the form [74]

$$g^{(2)}(t,t') = \frac{\langle a^\dagger(t)a^\dagger(t')a(t)a(t')\rangle}{\langle a^\dagger(t)a(t)\rangle\langle a^\dagger(t')a(t')\rangle} = \frac{\langle : I(t)I(t') :\rangle}{\langle I(t)\rangle\langle I(t')\rangle} ,$$

where :: denotes normal ordering and $I(t)$ the time-dependant intensity operator of mode a. In the special case of a stationary field, the equation

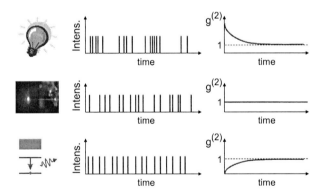

Figure 2.1: Photon counting events and corresponding second-order correlation functions for a classical state (light bulb), coherent light (laser), and a single-photon source, respectively.

Figure 2.2: Setup of a Hanbury-Brown and Twiss setup, including a 50 : 50 beam splitter and start- and stop-APD modules, respectively. Photon count coincidences are binned in a histogram.

simplifies to

$$g^{(2)}(\tau) = \frac{\langle : I(\tau)I(0) : \rangle}{\langle I(0) \rangle^2} , \qquad (2.1)$$

which relies on time differences τ only.

Evaluating equation 2.1 for different states of light reveals significant properties of the second-order correlation function [75]. For classical states, the Cauchy-Schwartz inequality demands $g^{(2)}(\tau) \leq g^{(2)}(0)$ for all τ. Since classical probability distributions of the electric field are positive definite, it can also be shown that $g^{(2)}(0) \geq 1$, thus $g^{(2)}(\tau) \geq 1$ follows. On the other hand, a single-mode quantum field with photon number variance $V(n)$ and mean $\bar{n}$ obeys

$$g^{(2)}(0) = 1 + \frac{V(n) - \bar{n}}{\bar{n}^2} .$$

For an eigenstate of the photon number operator (or Fock state), this yields $g^{(2)}(0) = 1 - 1/n$. Single-photon emission dominates if $g^{(2)}(0)$ drops to a value below $1/2$.

Figure 2.1 summarizes the behavior of $g^{(2)}$ for classical and quantum states. Classical states like the field emitted from a light bulb show an enhancement in the second-order correlation function near zero delay while single-photon sources exhibit a characteristic dip or *antibunching*. Sources with a Poisson distribution, like lasers well-above threshold, have a flat correlation function. Experimental characterization of a single-photon source involves monitoring its emission by a sensitive photo detector. In the visible, high detection efficiencies can be reached by an avalanche photo diode (APD). Unfortunately, their operation in Geiger mode requires a high bias voltage which needs to be rebuilt after each photon counting event. This leads to significant dead times of up to $t_{DT} = 50$ ns for a highly efficient APD [76] and inhibits the determination of correlations for $\tau < t_{DT}$.

The famous experiment by Hanbury-Brown and Twiss (HBT) [77], that aimed for the measurement of star diameters, offers a solution to advance towards shorter correlation times. Two APD modules detect the light behind the output ports of a 50 : 50 beam splitter (figure 2.2). The detection of a photon at APD 1 starts a coincidence counter which keeps running until APD 2 receives a click. These time intervals are then stored and binned in a histogram. An HBT setup not only avoids errors due to the detector dead time, but also allows to record asymmetric cross-correlation functions and to monitor the evolution of the histogram online. For details on the relationship between the theoretical $g^{(2)}$ function and an experimental HBT measurement, see [78].

2.2 Quantum Dots

Among the various quantum emitters, which have successfully generated non-classical light states in the past, quantum dots are of particular interest due to their optical stability and compatibility to today's semiconductor technology. The term "quantum dots" (QDs) often spans the wide group of colloidal nanoemitters [31, 79, 80] which have found applications in color glasses for centuries. In the following, however, it will solely be used for epitaxially grown or lithographically produced dots in the context of "coherent inclusions in a semiconductor matrix with truly zero-dimensional electronic properties" [81].

A bulk semiconductor is known for its quasi-continuous energy spectrum within certain band regions. Restricting its size in one or more dimensions leads to a quantization of the corresponding component of the electron wave vector and to a modification of the density of states (figure 2.3). Electron confinement can be realized in a semiconductor heterostructure by placing a material of low band gap ΔE_c inside a matrix that forms a higher surrounding potential. For quantum films, the density of states exhibits a staircase behavior, and the 3D confinement of QDs even results in single discrete energy levels. Thus, QDs are often referred to as "artificial atoms". While the actual geometry of the inclusions may vary – e.g., pyramids, cones, rings, and coupled structures have been realized [82] –, and intensive effort has been made to derive their level structure from *ab initio* calculations (e.g., [83, 84], for an introduction see [85]), the simple model of a cubic potential well is capable to explain this discretization qualitatively. The QD size must be chosen both to allow the existence of well-defined discrete levels by exceeding a minimum diameter $D_{min} = \pi\hbar/\sqrt{2m_e^*\Delta E_c}$ for an effective electron mass m_e^* and to avoid a substantial population of higher states by keeping their

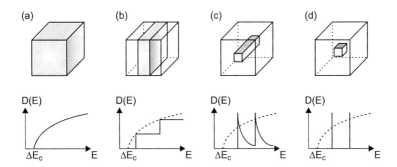

Figure 2.3: Modification of the density of states by consecutive electron confinement from bulk semiconductors to quantum dots. The well-known square-root dependance of the bulk (a) is altered to a stair-case behavior in quantum films (b) and further to discrete levels in quantum dots (c).

size small. Typical values for the most common systems are ∼ 12 nm for GaAs/AlGaAs and ∼ 20 nm for InAs/AlGaAs systems. A low temperature of operation can assure that the gross population is in the ground state.

The most obvious – and first realized – way to fabricate QDs, is the use of lithographic methods. Starting in the late 1980s, various techniques like optical, electron beam, and focused ion beam lithography, as well as scanning tunneling techniques were applied to quantum well patterning. It was difficult to reach sufficiently small structures around 10 nm, though. A new concept was the evolution of an initially two-dimensional growth into a three-dimensional corrugated growth, particularly in the presence of strain. Following the ideas of Stranski and Krastanov [86], self-organization of islands was observed in an InAs/GaAs superlattice [87] for the first time. It has been demonstrated using both molecular beam epitaxy (MBE) (e.g., [88], Ge on Si) and metal-organic chemical vapor deposition (MOCVD) (e.g., [89], InP/GaInP). While laser devices demand a high density of QDs, quantum information experiments rely on the probing of individual emitters. Epitaxial growth has to stop just above the critical wetting layer thickness which is about 1.5 monolayers for InAs/GaInAs and 1.8 monolayers for InP/GaInP heterostructures.

If a QD is not populated by charge carriers, it is considered to rest in its ground state (figure 2.4). Charge carriers may either be captured by the QD from the host matrix or created by optical or electrical excitation. Inside the QD, electrons and holes can form *excitons*, bound quasi-particles, which

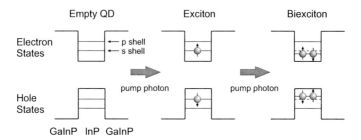

Figure 2.4: Energy levels in a QD populated by different excitations. The biexciton energy slightly differs from the excitonic level due to Coulomb interaction.

lowers the system energy due to Coulomb interaction. One distinguishes between the weak-confinement regime, where the QD exceeds the exciton Bohr radius and the exciton center-of-mass motion becomes quantized, and the strong-confinement regime in small QDs, where the kinetic energy contribution dominates due to size quantization [81]. The Pauli principle allows the population of each level by up to two carriers in different spin states. Due to a modified dispersion relation for holes in low-dimensional systems compared to bulk semiconductors, heavy holes with spin components $S_{hh}^z = \pm 3/2$ are energetically favored while electrons occupy $S_e^z = \mp 1/2$. This results in four possible combinations of uncharged exciton states with net spin components $S_{tot}^z = \{-2, -1, +1, +2\}$. Holes and electrons can recombine, emitting a single photon that carries away a net spin of ± 1, leaving two exciton dark states. Equivalently, two electron-hole pairs form a so-called *biexciton* that recombines in a cascade via one of the bright exciton levels to the ground state, with the biexciton-exciton transition energy slightly shifted from the exciton energy due to Coulomb interaction.

Various additional electronic states can be occupied in QDs, depending on the depth of the confining potential, e.g., charged excitons (*trions*) or more complex carrier combinations like *triexcitons* [90, 91]. The next chapter will provide an introduction to the generation of non-classical light states from QDs.

2.3 Micro-Photoluminescence Setup

The most fundamental physical resource in optical quantum information precessing is the single photon. Single quantum emitters like single QDs are

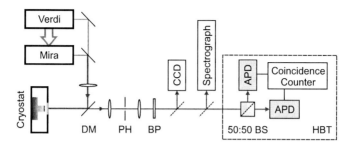

Figure 2.5: Schematic of the micro-photoluminescence setup. Pulsed or cw pump light is focused on a QD sample inside a liquid-He cryostat. The spontaneous emission passes a dichroic mirror (DM) and a band-pass (BP) for spectral filtering. A pinhole (PH) blocks stray light of nearby QDs. The emission is analyzed by a CCD camera, a spectrograph, or a HBT correlator.

naturally suited for its generation since only a single excited level is occupied at low excitation power and additional photon emission is delayed by the system's reexcitation time. Special effort and care need to be taken in order to reliably ensure the excitation of only one single emitter and to extract the generated photons efficiently. Here, the description will focus on optical excitation of QDs while details on electrical excitation are presented in chapter 3.

Low-density QD samples grown in Stranski-Krastanov mode exhibit typical dot densities of $10^8 - 10^{11}/\text{cm}^2$ and require sub-micron spot sizes to monitor single dots individually; micro-photoluminescence setups are capable to provide this resolution. Efficient detection must involve an objective of high numerical aperture and sufficient working distance to image a sample stored at low temperature inside a cryostat. High temporal resolution in coincidence measurements is achieved via an HBT correlator.

The total setup for single-photon generation is depicted in figure 2.5. In order to limit line broadening due to phonon-phonon or electron-phonon interaction, the QDs must be stored under cryogenic conditions. A continuous-flow liquid Helium cryostat (Konti-Cryostat-Mikro, CryoVac) cools the sample down to 4.2 K and minimizes absolute temperature fluctuations to about 0.1 K. A built-in two-axes step motor allows lateral alignment of the sample with a travel range of 5 mm and a precision of 0.1 μm. The sample is attached to the cooling finger of the cryostat via heat-conductive paste that reduces stress inside the sample during cooling cycles and minimizes long-term mechanical drifts. Thus, stable imaging of a single QD is achieved for

15

over more than one hour. A 0.5 mm thick window provides optical access. The QDs may be illuminated either continuous-wave (cw) or by a pulsed laser depending on the experiment of interest. They must be excited at or above the exciton energy. Excitation of slightly higher continuum states is followed by rapid non-radiative decay to the exciton level. The cw source is a frequency-doubled Nd:YVO$_4$ laser pumped by a laser diode array (Verdi, Coherent). It provides 10 mW − 10 W of linearly polarized output power at a wavelength of 532 nm with a specified linewidth of 5 MHz. This linewidth can be considered a δ-distribution compared to exciton resonances of a few GHz. For pulsed excitation, the cw source pumps a mode-locked Ti:Sa laser (Mira, Coherent) that is tunable between 750 − 930 nm. A pulse width of 390 fs has been measured [78] at a repetition rate of 76 MHz, corresponding to a Fourier limited spectral width of 1.38 nm. In order to obtain a sufficient excitation photon energy, the Ti:Sa output is frequency-doubled in a beta barium borate crystal yielding up to 50 mW mean power of blue light.

An objective of high numerical aperture (NA) is needed to achieve a tight focus at the sample as well as large collection efficiency for the fluorescence; it represents the critical part in any micro-photoluminescence setup. The resolution Δx of conventional microscopes is limited by the Rayleigh criterion [92]

$$\Delta x = 0.61 \frac{\lambda}{\mathrm{NA}} \,.$$

While the exciton wavelength λ is fixed, the NA is a valid parameter to increase the resolution. Therefore, a telescope expands the excitation beam diameter to 1 cm before it enters the objective. Unfortunately, the cryostat geometry sets a lower limit to the working distance and avoids the use of immersion oil. A well-known technique to improve the resolution is scanning confocal microscopy [93]. Here, the theoretical limit is set by

$$\Delta x = 0.44 \frac{\lambda}{\mathrm{NA}} \,.$$

Detection through a pinhole illuminates only a small sample volume which inhibits wide-field imaging if needed. An additional lens on a flip mount, however, that focusses the excitation light into the back focal plane of the objective, yields full access to both high resolution and wide-field images. Details on this setup can be found in [78]. The selected objective (63× Achroplan, Zeiss) offers a high NA = 0.75 at a comparably large working distance of 1.3 mm. It is corrected for abberations due to glass cover slips or the cryostat window. The objective is mounted to a piezo tube to account for the variable heights of different samples. The maximum spatial resolution of the setup was measured to $\Delta x_{\mathrm{exp}} = 0.53$ μm [78]. In order to achieve a higher setup transmission, a 1 μm pinhole was used, though.

The QD emission passes a dichroic mirror for further analysis. A CCD camera (Hamamatsu 1394 ORCA-ER, Hamamatsu Photonics) monitors the sample and helps to focus on individual dots. In order to select a dot with a strong exciton line, a 0.5-m grating spectrograph in combination with a liquid-Nitrogen-cooled CCD camera is used (Acton 500i, Roper Scientific). Its spectral resolution was measured to be $\sim$ 100 μeV, corresponding to $\sim$ 40 pm at 700 nm.

Finally, two APD modules in HBT configuration are used to characterize the photon statistics with a time resolution beyond the detector dead time. Each APD (SPCM-AQR-14, PerkinElmer) shows electrical dark count rates below 100 counts/s, and background rates of 100 and 150 counts/s, respectively, have been measured for free-beam operation inside a black box with small apertures. This APD is designed for photon detection in the visible reaching its maximum quantum efficiency of 0.70 at $\sim$ 700 nm. Cross-talk between the modules, as reported in [94], was minimized by additional pinholes in front of each APD. In order to record correlation functions, the TTL output pulses of the modules (width: $\sim$ 50 ns) are fed into a coincidence counter (TimeHarp 100, PicoQuant). This PCI card allows to measure time differences between start and stop events down to 40 ps. The actual resolution of the HBT correlator does not exceed 800 ps, though, due to the timing jitter of the APD modules.

A careful discussion of this experimental setup must conclude with an estimation of the excitation and collection efficiency. At low temperature, thermal carrier escape out of the dot potential is negligible. Thus, the excitation efficiency is close to unity if the pump drives the exciton transition in saturation. Besides, no significant non-radiative decay channel exists for the exciton level [95]. The collection efficiency is the product of extraction efficiency from the QD sample into the objective, setup transmission, and detection efficiency. The extraction mainly depends on the refractive index of the sample which governs both the angle of total reflection and the collection solid angle of the objective. A detailed evaluation [78] yields a value of < 0.01 for InP ($n = 3.5$). Micro-resonators around the dot [96–98], solid-immersion lenses [99], or an additional mirror layer [100] can enhance this efficiency. Together with a setup transmission of $0.5 - 0.6$ and a detection efficiency of 0.7 at 700 nm, the collection efficiency, and therefore the total single-photon generation efficiency, is estimated to $0.001 - 0.002$. Thus, count rates of $75 \times 10^3 - 150 \times 10^3$ counts/s are expected at pulsed excitation with a Ti:Sa laser of 80 MHz repetition rate.

2.4   Single Photons from InP Quantum Dots

Single-photon states have been generated from various semiconductor heterosystems for many years. First realizations were achieved with InAs QDs in GaAs [20, 101]. III-V compounds had the advantage of easier fabrication while II-VI systems show especially small lifetimes which are promising for high photon generation rates. Depending on the band gap of the chosen heterostructure, miscellaneous emission wavelengths can be realized. For long-distance applications, the absorption and dispersion minimum of optical fibers (0.2 dB/km at 1550 nm [102]) is favorable, but efficient single-photon detectors beyond 1 μm are still subject of intense research, suffering from low efficiencies and bad signal-to-noise ratio. Using super-conducting single-photon detectors [103], a quantum efficiency QE $\sim$ 0.6 at 1.55 μm has been achieved [104], and even photon number resolution is possible [105]. The best quantum efficiency of commercial photon counters today was measured around 700 nm (QE $\sim$ 0.7 [76]) which fits exciton transitions of InP/GaInP QDs. Thus, InP/GaInP QDs are well-suited for free-beam single-photon experiments. Studies on InP/GaInP [89, 106] showed a bimodal size distribution with the smaller dots (height < 5 nm, width $\sim$ 40 nm) emitting at the desired wavelength around 690 nm.

Single-photon generation from the QD sample under investigation has been reported in [100] for the first time. The sample structure is depicted in figure 2.6. The heterostructure was grown by MOCVD on a GaAs wafer. On top of a GaAs buffer, 300 nm of lattice-matched GaInP was evaporated. In Stranski-Krastanov mode, 1.9 monolayers of InP formed QDs with a small-dot density of 10^8 cm^{-2} which were capped by another 100 nm of GaInP to confine the dot in all spatial dimensions and to avoid chemical adsorption. In a next step, a 200 nm Al film was deposited in order to increase the photon collection efficiency. Finally, the sample was glued upside down onto a Silicon substrate, and the GaAs wafer was removed by selective wet etching. Before the sample can be used in subsequent quantum information experi-

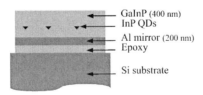

GaInP (400 nm)
InP QDs

Al mirror (200 nm)
Epoxy

Si substrate

Figure 2.6: Structure of the InP/GaInP sample under investigation. An additional Al layer has been deposited for enhanced collection efficiency [100].

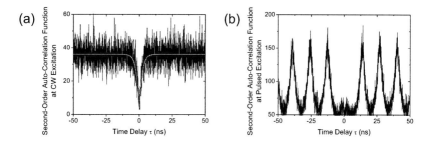

Figure 2.7: Strong suppression of two-photon events proven by (a) cw and (b) pulsed intensity auto-correlation measurements on a single quantum dot.

ments, the single-photon character of its light emission needs to be proven. QDs of narrow spectral width were preselected by wide-field imaging and by monitoring their spontaneous emission behind a 1-nm wide band-pass filter. The sample was kept at a temperature of < 10 K in order to minimize line broadening. The setup was then switched to confocal mode in order to confirm the narrow emission of a particular dot by spectroscopic measurements. A linear dependance of the spontaneous emission rate on pump power further raises evidence of an exciton line while bi- and triexcitons are known to show parabolic and cubic behavior, respectively. These higher excitations can be suppressed in a sufficiently low excitation regime and by using a narrow band-pass filter (figure 2.5).

Figure 2.7a depicts measurements of the second-order correlation function on the selected dot at both continuous and pulsed excitation which furnishes ultimate proof of single-photon emission. The cw graph shows strong antibunching down to a value $g^{(2)}(0) = 0.11$ for the normalized function that clearly falls below the single-photon limit of $1/2$. This value, however, can totally be attributed to the finite time resolution of the HBT setup. A fit function was determined by convoluting the theoretical normalized autocorrelation function at low excitation $g^{(2)}(\tau) = 1 - \exp\left(-|\tau|/\tau_{exc}\right)$ with a Gaussian distribution of width $w = 800$ ps that accounts for the limited HBT resolution (fit curve in figure 2.7a). Exciton lifetimes $\tau_{exc} \approx 2$ ns have been measured in a previous independent experiment [78]. Recalculating the $g^{(2)}$ fit function with a HBT time resolution $w \ll \tau_{exc}$ yields $g^{(2)}(0) = 0.003$. In order to operate the setup as a deterministic single-photon source, providing single photons *on demand*, pulsed excitation is inevitable. The QD selected for measurement shows a strong suppression of the central second-order correlation peak (figure 2.7b). This measurement proves single-photon emission; i.e., the probability of two-photon events between two successive pump pulses is substantially reduced beyond the limit of a two-photon Fock state.

2.5 Deutsch-Jozsa Algorithm

The InP/GaInP single-photon source may now show its applicability to tasks in photonic quantum information processing like all-optical quantum computation. The Deutsch-Jozsa algorithm is often considered a toy model for more complex quantum computation. It efficiently determines the constant or balanced nature of a boolean function – a task that has not found its way into any practical application to date, though.

In classical information science, a boolean function or gate links a register of input bits and an output bit by a certain directive: $f : \{0,1\}^n \rightarrow \{0,1\}$. We now focus on two special classes, the *constant* and *balanced* functions:

$$f \text{ constant}: \quad f(x) = 0 \text{ or } 1 \quad \forall x = (x_1 \ldots x_n) \text{ with } x_i = 0 \text{ or } 1$$
$$f \text{ balanced}: \quad f(x) = 0 \qquad \text{for exactly one half of the domain}$$

In a classical computation, up to $2^{n-1} + 1$ evaluations of the function f may be necessary to distinguish between these two classes. Quantum mechanics, however, performs this distinction with a single evaluation.

This quite formal approach is sometimes visualized by a coin tossing game (figure 2.8). Imagine a competition between Alice and Bob – the two generic characters of quantum information science –, where Bob may be dealing real coins (with different sides) or trick coins (with equal sides). How many sides of these coins will Alice need to check before she can be sure he is not counterfeiting? Again, the realistic classical answer to this question is $2^{n-1} + 1$.

Figure 2.8: Quantum mechanics can help Alice in a coin tossing game if Bob tries to cheat on her.

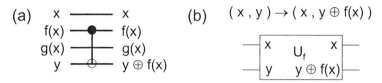

Figure 2.9: (a) Application of the CNOT gate, (b) block diagram for the unitary transform U_f.

In a quantum world with interfering coins, however, Alice can save a lot of time.

Since linear-optics quantum computation relies on unitary transformations, f must be transcribed in terms of reversible gates. A basic theorem of quantum information processing [107] allows the expansion of classical gates into universal reversible gates, e.g., the Toffoli gate. However, ancilla or garbage qubits may be needed as the price to pay for reversibility:

$$(x, 0, 0) \rightarrow (x, f(x), g(x)) .$$

The ancilla and garbage qubit, respectively, must be initialized as 0 or 1 depending on the implemented function f. A fourth qubit is required to avoid the accumulation of qubits in states sensitive to the computation and to reset the ancilla qubits:

$$(x, 0, 0, y) \rightarrow (x, f(x), g(x), y) .$$

Finally, $f(x)$ is transferred onto the fourth qubit using a reversible CNOT as depicted by figure 2.9(a)

$$(x, 0, 0, y) \rightarrow (x, f(x), g(x), y \oplus f(x)) ,$$

and the inverse operation (excluding CNOT) on qubits 1–3 yields

$$(x, 0, 0, y) \rightarrow (x, 0, 0, y \oplus f(x)) .$$

The block diagram of the corresponding unitary transform U_f is depicted in figure 2.9(b).

We can now proceed to the actual quantum circuit which performs the Deutsch-Jozsa algorithm (figure 2.10). The implementation starts with the initialization of the $(n + 1)$-qubit register into

$$|\psi_0\rangle = |0\rangle^{\otimes n}|1\rangle .$$

The initial Hadamard gates modify this state to

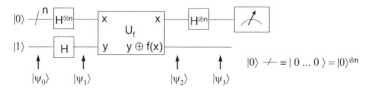

Figure 2.10: Quantum circuit for the n-qubit Deutsch-Jozsa algorithm.

$$|\psi_1\rangle = \left(H^{\otimes n} \otimes H\right)|\psi_0\rangle = \left(\frac{1}{\sqrt{2}}\right)^{n+1} \sum_{x \in \{0,1\}^n} |x\rangle \otimes (|0\rangle - |1\rangle)$$

and generate a uniform superposition of all $|x\rangle$. The unitary transform U_f represents the heart of the algorithm and leaves the register in the state

$$|\psi_2\rangle = U_f|\psi_1\rangle = \left(\frac{1}{\sqrt{2}}\right)^{n+1} \sum_{x \in \{0,1\}^n} (-1)^{f(x)} |x\rangle \otimes (|0\rangle - |1\rangle) \ .$$

U_f evaluates $f(x)$ for all 2^n n-qubit states which may be considered a proof for the remarkable capability of this formalism to perform parallel computation. A measurement on the first n qubits yields $f(x)$ for a single $|x\rangle = (x_1 x_2 \ldots x_n)$ only, though, and an additional application of $H^{\otimes n}$ on the first n qubits is required to reveal the actual speed-up:

$$
\begin{aligned}
|\psi_3\rangle &= \left(H^{\otimes n} \otimes \mathbb{I}\right)|\psi_2\rangle \\
&= \left(\frac{1}{2}\right)^n \sum_{y \in \{0,1\}^n} \sum_{x \in \{0,1\}^n} (-1)^{x \cdot y + f(x)} |y\rangle \otimes \frac{1}{\sqrt{2}} (|0\rangle - |1\rangle) \ .
\end{aligned}
$$

Let us focus on the summand $|y\rangle = |00 \ldots 0\rangle = |0\rangle^{\otimes n} = |0\rangle$. Since $x \cdot y = 0$, its amplitude reads

$$A_{|0\rangle^{\otimes n}} = \left(\frac{1}{2}\right)^n \sum_{x \in \{0,1\}^n} (-1)^{0+f(x)} = \left(\frac{1}{2}\right)^n \sum_{x \in \{0,1\}^n} (-1)^{f(x)}$$

f **constant:**   In this case, we have

$$A_{|0\rangle^{\otimes n}} = \left(\frac{1}{2}\right)^n \sum_{x \in \{0,1\}^n} (-1)^0 = +1 \ \text{and} \ A_{|0\rangle^{\otimes n}} = \left(\frac{1}{2}\right)^n \sum_{x \in \{0,1\}^n} (-1)^1 = -1 \ ,$$

respectively. A measurement on the first n qubits thus leads to the state $|0\rangle^{\otimes n}$ with certainty. All qubits rest in state $|0\rangle^{\otimes n}$.

f **balanced:**   Here, the amplitude is

$$A_{|0\rangle^{\otimes n}} = \frac{1}{2} \left(\sum_{x\in\{0,1\}^n} (-1)^0 + \sum_{x\in\{0,1\}^n} (-1)^1 \right) = 0 \, .$$

A measurement on the first n qubits thus never leads to the state $|0\rangle^{\otimes n}$. In summary, if a measurement on the first n qubits yields at least a single "1", then f is balanced, otherwise, f is constant. The algorithm solely works for the limits of constant and balanced functions since only then $A_{|0\rangle^{\otimes n}}$ is strictly "0" or "±1".

2.6   Deutsch-Jozsa Implementation

The Deutsch-Jozsa algorithm has been implemented in a number of physical systems to show their principal applicability to quantum computational tasks. Nuclear magnetic resonance [108, 109], solid-state [110], and ion-trap [111] realizations have already been demonstrated. Previous all-optical experiments were restricted to emulate the Deutsch-Jozsa algorithm with attenuated classical laser pulses [112–114]. However, a deterministic single-photon source is essential for many applications of LOQC, providing single photons *on demand*, since a two-photon admixture may disturb single-photon interference. Triggered emission from single QDs represents such a true deterministic single-photon source and will be applied in the following scheme [115].

In its simplest realization, the Deutsch-Jozsa algorithm requires one qubit to store the "query" and a single qubit to store the "answer". In the following implementation, the first qubit is encoded in the photon path, namely, the two spatial modes a and b of a Mach-Zehnder interferometer (figure 2.11). In this dual-rail representation, the logical state $|0\rangle_L$ ($|1\rangle_L$) is associated with the physical state $|a\rangle$ ($|b\rangle$) which corresponds to the situation "one photon in mode a (b)". The second qubit is implemented via the photon's polarization state $|H\rangle$ ($|V\rangle$), its generic degree of freedom for qubit encoding.

Adapting the algorithm: The two qubits are initialized in the state

$$|\psi_1\rangle = \frac{1}{2} \left(|0\rangle_L + |1\rangle_L \right)_x \otimes \left(|0\rangle_L - |1\rangle_L \right)_y$$

where x and y indicate query and answer qubit, respectively. The initial state $|\psi_1\rangle$ is realized by sending a photon into a linear superposition of the two spatial modes using a non-polarizing $50 : 50$ beam splitter after preparing it

23

in a superposition of horizontal ($|H\rangle$) and vertical ($|V\rangle$) polarization. This results in

$$|\psi_1\rangle = -\frac{1}{2}\left(|a\rangle + i|b\rangle\right)_x \otimes \left(|H\rangle - |V\rangle\right)_y$$

where the relative phase $\exp(i\pi/2) = i$ in the first qubit accounts for a reflection at the beam splitter that does not affect the overall performance of the algorithm.

In a next step, a unitary transform

$$|\psi_1\rangle \rightarrow |\psi_2\rangle = \pm\frac{1}{2}\left[(-1)^{f(a)}|a\rangle + i\,(-1)^{f(b)}|b\rangle\right]_x \otimes \left(|H\rangle - |V\rangle\right)_y$$

is realized by selectively adding a half-wave plate (HWP) to each mode. The four possible setups represent the functions $f(a) = f(b)$ (constant functions) and $f(a) \neq f(b)$ (balanced functions), corresponding to two equal- and two different-sided coins, respectively. The transformation above will be achieved if the value $f(x) = 0$ is assigned to "wave plate installed in arm $x \in \{a, b\}$" and the value $f(x) = 1$ to "no wave plate in arm x". Recombining the two interferometer arms on a second non-polarizing $50 : 50$ beam splitter implements a final Hadamard gate on the first qubit, yielding the state

$$|\psi_2\rangle = \begin{cases} \pm\frac{i}{\sqrt{2}}|b\rangle_x \otimes \left(|H\rangle - |V\rangle\right)_y & (f \text{ constant}) \\ \pm\frac{1}{\sqrt{2}}|a\rangle_x \otimes \left(|H\rangle - |V\rangle\right)_y & (f \text{ balanced}) \end{cases}.$$

Experimental realization: The experimental setup resembles a classical Mach-Zehnder interferometer (figure 2.11). Constructive (destructive) interference in the extension of mode a indicates a balanced (constant) function, i.e., one and only one wave plate (two or none wave plates) in the beam paths. The quantum character of this computation becomes obvious by the fact that it uses single-particle interference and that a single detection event is sufficient to test the oracle function f.

The single-photon input state is generated by spontaneous emission from a low-density QD sample as described in section 2.3. All measurements are performed at both cw and pulsed QD excitation. The photoluminescence from the exciton transition of a single dot is fed into the interferometer setup with an overall collection efficiency of about 10^{-3}. As the QD emission is unpolarized, a polarizer initializes the transmitted photons in the state $|+\pi/4\rangle \equiv -1/\sqrt{2}\,(|H\rangle - |V\rangle)$.

The photons enter the Mach-Zehnder interferometer (figure 2.11) that is actively stabilized to the interference signal of a He-Ne laser which propagates about 1 cm above the single-photon field. A mirror mounted on a piezoelectric stage compensates for drifts of the optical path difference and keeps its

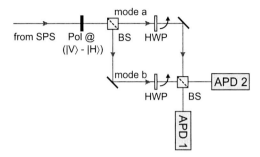

Figure 2.11: Experimental setup: Linearly polarized single photons enter an interferometer via a non-polarizing 50 : 50 beam splitter (BS). The two balanced and two constant functions are realized via conditionally adding half-wave plates (HWP) into the arms, and a second BS performs a final Hadamard gate. Two single-photon detectors evaluate the algorithm output and confirm single-photon emission.

deviation below a fraction of the wavelength via a servo system. The two interferometer outputs are monitored with one APD each.

Measurements: Figure 2.12 depicts the normalized count rate I_{APD1} at APD 1 (complementary count rate I_{APD2} at APD 2), corresponding to the four possible HWP combinations in the interferometer arms, i.e., the four possible functions. The count rates show a mean fringe contrast

$$C = |I_{APD1} - I_{APD2}| / (I_{APD1} + I_{APD2}) \approx 0.75$$

($C \approx 0.64$ at cw excitation) and reveal a clear intensity change at each APD when the HWP combination is switched between a balanced and a constant function. This contrast can be explained by the limited coherence length of the photons and a tiny mismatch of the interferometer arm lengths. Previous measurements showed that an interferometer path difference of only 100 μm results in a reduction of interference contrast down to 0.80 [116]. The visibility

$$V = \frac{\left| I_{APD1/2,\text{bal}} - I_{APD1/2,\text{const}} \right|}{I_{APD1/2,\text{bal}} + I_{APD1/2,\text{const}}}$$

represents a measure for the success probability of our computation. Mean values of $V \approx 0.79$ and $V \approx 0.71$ have been achieved for APD 1 and APD 2, respectively. Imperfections can be attributed to slight inaccuracies in the relative path lengths of the interferometer and background noise. The superior

25

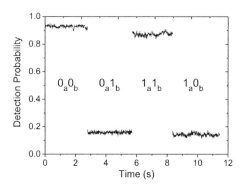

Figure 2.12: Count rate at APD1 corresponding to the realization of the four oracle outcomes. The four functions are denoted by the number of HWPs in mode a and b, respectively. The normalized count rate yields a visibility of about 0.79.

value of the APD 1 signal results from the lower background created by He-Ne stray light in this detector channel. Analogous measurements have been performed at cw excitation, yielding visibilities around $V \approx 0.67$ for both APD 1 and APD 2, presumably due to a lower stability of the interferometer during the cw measurement.

A strong contrast dependence on the angle of the initial polarization testifies that this degree of freedom corresponds to the answer qubit in the formulation of the Deutsch-Jozsa algorithm. For instance, an answer qubit initialized in the state $|V\rangle$ allows no definite prediction of the constant or balanced nature of the applied function. Thus, a correct initialization of the polarization is crucial. In figure 2.13, the dependence of the APD signals is displayed for different HWP combinations at pulsed excitation. In the case of constant functions, the contrast is independent of the initial polarization, as the transformation induced by the HWP is symmetric in both interferometer arms. The degradation of contrast at larger angles can be attributed to a thermal drift of the QD sample during the measurement. The situation is different for the balanced functions when the HWP rotates the polarization in only one arm, dependent on the angle between initial polarization and the optical axis of the HWP. At $\phi = 0°$, the interference contrast vanishes completely. The fit in figure 2.13(b) shows the expected shape $c\,[d + \sin^2{(\phi\,\pi/180° + \pi/4)}]$ for the configuration "wave plate in mode b" where $c \approx 0.75$ and $d \approx 0.16$ account for a limited overall contrast and for an offset due to dark counts,

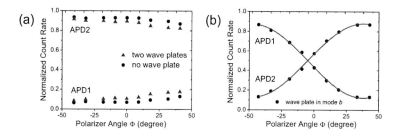

Figure 2.13: Influence of the initial single-photon polarization on the normalized APD count rates. (a) Constant visibility characteristics after implementation of the two constant functions, (b) visibility following a squared sine in case of the two balanced functions. The fit already accounts for imperfections like dark counts and limited contrast. For better readability, only the case "wave plate in mode b" is fitted.

respectively ($c \approx 0.71$ and $d \approx 0.27$ for "wave plate in mode a"). Similar results have been obtained at cw excitation.

Error correction: The previous results demonstrate triggered operation of a quantum algorithm using a deterministic single-photon source. However, the treatment of errors during computation has become most relevant for quantum algorithms [117, 118]. The concept of decoherence-free subspaces [119, 120] opens a way to make quantum computation passively stable by encoding information in certain parts of a Hilbert space that are not affected by the noise operators. A variation of the experimental setup (figure 2.14) enables the implementation of these ideas in a triggered quantum algorithm on the single-photon level. It is possible to encode the qubits in a way that they are no longer affected by phase noise which is generally the primary error source in optical interferometric setups. By properly separating and merging horizontal and vertical polarization, phase noise $\Delta\theta$ in the central interferometer does not cause perturbations of the algorithm. In this particular experimental realization, two noise resistant combinations are

$$|H\rangle_x \otimes |a\rangle_y - |V\rangle_x \otimes |b\rangle_y \quad \text{and} \quad |H\rangle_x \otimes |b\rangle_y - |V\rangle_x \otimes |a\rangle_y$$

This resistance can be understood as follows: The photon state

$$\left|\tilde{\psi}_1\right\rangle = \pm \frac{1}{2} \left(|a\rangle \pm i|b\rangle\right)_x \otimes \left(|H\rangle - |V\rangle\right)_y$$

27

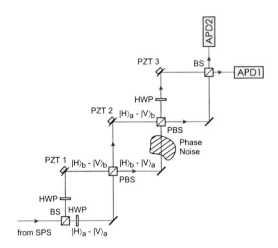

Figure 2.14: Modified experimental setup for noise resistant measurements (PBS: polarizing beam splitters, BS: non-polarizing 50 : 50 beam splitters). PZT 1 stabilizes the first part of the interferometer as described before. PZT 2 and 3 selectively modulate the various arm lengths of the interferometer.

(for constant and balanced functions f, respectively) before the first polarizing beam splitter (PBS) is transformed into

$$\left|\tilde{\psi}_2\right\rangle = \pm\frac{1}{2}\left\{[\exp\left(i\Delta\theta\right)|a\rangle \pm i|b\rangle]_x \otimes |H\rangle_y + [|a\rangle \pm i\exp\left(i\Delta\theta\right)|b\rangle]_x \otimes |V\rangle_y\right\}$$

after the second. The final state at the detectors then reads

$$\left|\tilde{\psi}_f\right\rangle = \begin{cases} \pm\frac{i}{\sqrt{2}}|b\rangle_x \otimes [\exp\left(i\Delta\theta\right)|H\rangle - |V\rangle]_y & (f \text{ constant}) \\ \pm\frac{1}{\sqrt{2}}|a\rangle_x \otimes [\exp\left(i\Delta\theta\right)|H\rangle - |V\rangle]_y & (f \text{ balanced}) \end{cases}.$$

Obviously, theory suggests that the induced phase noise has no impact on the detection events since APD modules are insensitive to polarization. In this experiment, piezo (PZT) modulation simulates this noise. A variation of the central arms (noise resistant combinations) by PZT 2 does not affect the count rates while phase noise, induced by PZT 3 for comparison, changes the APD signals substantially (figure 2.15). Beside phase noise, this setup is also insensitive to polarization bit flips, but experimental proof has been postponed to future measurements.

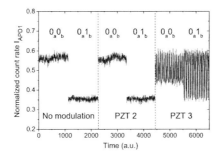

Figure 2.15: Phase noise simulated by PZT 2 does not change the detection events at APD 1 while path variations due to PZT 3 imprint strong perturbations in the count rates.

2.7 Discussion

The Deutsch-Jozsa algorithm was experimentally realized with linear optics using a deterministic single-photon source. Reliable distinction between the two classes of constant and balanced functions has been achieved with a visibility $V = 0.79$ at pulsed excitation. This value is interpreted as the success probability of this computation. Using an extended experimental setup, the robustness of this implementation against phase noise could be proven when information is properly encoded in an unaffected superposition of polarization states. This experiment shows that photons on demand from a single QD are capable of reliable triggered quantum computation. Therefore, this single-photon source based on spontaneous emission from a single QD exhibits key properties demanded by the KLM scheme proposed in [11] and proves the applicability of QD emission for LOQC.

A main challenge to be tackled in further experiments remains the generation of indistinguishable photons from at least two different sources in order to increase the number of qubits. Fourier limitation of single photons emitted by independent single molecules has recently been reported in [121]. In addition, the integration of entangled photons – first proposed in [122] and demonstrated in [123, 124] – will make more extended quantum computational tasks possible. Progress towards the realization of single-photon sources for more complex quantum networks will be the subject of part II of this thesis.

Chapter 3

The Ultimate Limit of an LED

"The principles of physics,
as far as I can see,
do not speak against the possibility,
of maneuvering things atom by atom."
– Richard Phillips Feynman

Coherent control of quantum systems is inevitable for the success of quantum information processing, and there has been tremendous progress in this field over the last decades. But still, implementations do not exceed few qubit realizations due to the destructive influence of decoherence. Thus, the field of experimental quantum information processing is mainly limited to proof of principle experiments with real-life applications far out of reach. The only exception to date is quantum cryptography [8, 125] which allows the secret distribution of a key between distant parties. It has demonstrated its potential, e.g., for safe bank transactions [126], and even commercial systems are available (idQuantique, Toshiba). Simple bipartite schemes can be realized with single qubits, and single photons are the system of choice for long-distance implementations. Thus, efficient, easy-to-handle, and compact all-solid state single-photon sources are key devices in the field of quantum communication. However, a true deterministic single-photon source is inevitable since security in long distance communication is crucially limited by contributions of higher photon numbers and is also a requisite for realizations of all-optical quantum computation [11, 115, 127]. Unfortunately, true single-photon sources based on single quantum emitters – contrary to attenuated laser pulses [128] – usually require large-scale laser systems and often cryogenic setups, as well.

The solution for QD implementations is electrical pumping, offering integrated and packaged light sources with non-classical emission characteristics. Early schemes utilized a simultaneous Coulomb blockade for electrons and holes in a semiconductor triple quantum well nanostructure [129]. Other realizations embedded single self-organized QDs in pin-diode structures [36, 130–133]. Their emission can be enhanced and directed into distinctive modes by the growth of Bragg reflectors [134]. But the desired ultimate control in integrated ready-to-go single-photon sources for quantum communication includes the deterministic injection of a single electron and a single hole into a QD in order to avoid the need for spectral or spatial filtering of the emission. It has been shown [135] that a single QD in a pin-diode can selectively be pumped in order to obtain a pure emission spectrum with only a single exciton line. Its application to quantum information processing requires the proof of non-classical light emission from this device.

3.1 Sample Fabrication

A promising approach for single-photon generation based on electrical pumping uses a micrometer-size aluminum oxide aperture to restrict the current flow to a single dot. Thus, electrical excitation of more than one dot is significantly suppressed, and non-classical statistics of the electroluminescence could be measured [131, 132]. But those experiments did not finally prove the injection of a single electron and a single hole into the pin-junction that would generate light of sub-Poissonian statistics without external filtering. Here, a different device is used for non-classical light generation (figure 3.1 [135, 137]). It is grown on semi-insulating (100) epi-ready GaAs

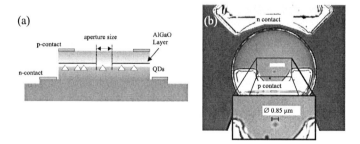

Figure 3.1: Schematic cross section (a) and microscope image (b) of the device structure [136]. For details see text.

substrates using a Riber-32P MBE system. The light emitting diode (LED) consists of an undoped GaAs layer with InAs QDs of low density inserted, a 60 nm thick aperture layer of high aluminum content AlGaAs, and p- and n-type GaAs electrical contact layers. The low QD density of 10^8 cm^{-2} was obtained in the Stranski-Krastanow mode by deposition of 1.8 ML of InAs. Cylindrical mesas were processed by inductively coupled plasma reactive ion etching, and selective oxidation of the high aluminum content AlGaAs layers led to sub-micrometer-size oxide current apertures. Subsequent Si$_3$N$_4$ deposition allowed Au/Pt/Ti and Au/Au-Ge/Ni metallization to form p- and n-contacts, respectively.

3.2 Measurements

In order to probe the electrical excitation of only a single QD, the electroluminescence spectrum was measured at an injection current of 870 pA and a bias voltage of 1.65 V (figure 3.2a). It reveals just one single line, and emission of the wetting layer is completely suppressed. For experiments related to entangled pair generation using biexciton-exciton decays [122], emission from an uncharged dot is requisite. Via high-resolution spectroscopy, a fine structure splitting of 55 μeV due to electron-hole exchange interaction could be determined. The existence of a splitting proves the electroluminescence to originate from an exciton rather than a trion state and will allow further

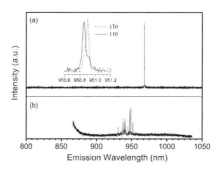

Figure 3.2: (a) Electroluminescence spectrum at a current of 870 pA and 1.65 V bias voltage (b) Micro-photoluminescence spectrum using a laser spot size of 2 μm. The inset shows the polarized fine structure splitting of the exciton line [136].

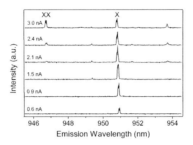

Figure 3.3: Electroluminescence spectra measured at increasing injection current. At higher current, the biexciton line (XX) appears beside the exciton emission (X). For better visualization, an offset has been added to each spectrum.

experiments towards entanglement generation with this device. Still, other dots may be excited and their electroluminescence be blocked by a shadow mask. The micro-photoluminescence of a few dots (figure 3.2b) at 10 K, however, exhibited a set of several discrete lines. This proves that the oxide aperture above the dot under study is transparent for near-infrared light, and the absence of light emission other than from the exciton decay in the electroluminescence spectrum clearly demonstrates the pumping of indeed only a single QD.

In figure 3.3, six electroluminescence spectra are depicted to further characterize the emission with increasing injection current. At around 1 nA, corresponding to an injection of 5 electrons/ns, the exciton emission (X) saturates. With a typical exciton lifetime of 1 ns [138–140] and an internal quantum efficiency of about unity [20, 81, 141], this gives a remarkable injection efficiency of about 0.2 since the capture time for carriers is in the picosecond regime. This is two orders of magnitude better than in previously reported structures [36, 130–132]. At higher currents, two additional lines appear. They both originate from the same dot since they obey the same spectral jitter. The high-energy peak is assigned to the biexciton decay (XX) due to its super-linear dependence on the injection current. The third line is unpolarized and does not show any splitting; thus, it can be attributed to the trion. As already mentioned, the emission from uncharged dots paves the way for on-demand generation of entangled photon pairs from this device which has not been possible in earlier realizations [132].

Given these extremely clean emission features and excellent carrier control, the applicability of this device as a single-photon source needs to be further

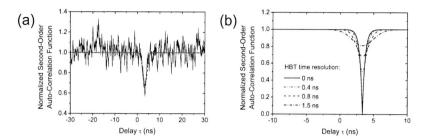

Figure 3.4: (a) Auto-correlation measurement at cw current injection (0.9 nA, 1.65 V) at 10 K. No spectral filtering was used to isolate a single transition in a single quantum dot. Dashed line: Fit function as described in the text. (b) Impact of the HBT time resolution on the measured correlation function.

exploited. Non-classical photon statistics is characterized by the second-order auto-correlation function via a HBT setup. In contrast to all previous HBT setups related to QD experiments, no spatial or spectral filtering of the exciton line was needed – except a 10 nm wide band pass filter centered at 953 nm in front of one APD to avoid cross-talk between the detectors.

The result of the auto-correlation measurement is depicted in figure 3.4(a). In order to interpret the experimental data, the rate model described in [78] was used. While an ideal single-photon source shows an antibunching dip of the second-order correlation function down to zero, the limited time resolution of the HBT setup must be taken into account. Thus, the ideal function $g^{(2)}(\tau) = 1 - \exp(-|\tau|/\Delta t)$, where Δt accounts for excitation and exciton decay rates, was convoluted with a Gaussian of 800 ps width which yields a dip depth of around 0.65. This time resolution was independently determined by an auto-correlation measurement of ultra-short laser pulses. Additional background, that may also result in a non-zero contribution to the antibunching dip can be excluded due to the excitation of only a single dot (as discussed above) and thus the purity of the spectrum. APD cross-talk was also eliminated, and it was checked that the APD dark count rate did not exceed the specified level of 100 Hz when blocking the QD emission. The convoluted fit shown in figure 3.4(a) agrees well with the experimental data. Figure 3.4(b) shows the impact of the time resolution on the $g^{(2)}$ function for clarity. Thus, the non-zero dip can be explained by the limited time resolution of the HBT setup, and the measurement proves that this QD LED represents indeed an ideal electrically pumped single-photon source.

3.3 Discussion

The measurements characterize this QD LED with sub-micrometer-size aperture as a highly efficient light source that allows the controlled injection of a single electron and a single hole into a single QD. Photons from uncharged dots can be extracted in well-defined polarization modes, and photon statistics exhibits strong antibunching. Together with its high efficiency and unmatched spectral purity, the presented structure is predestined for the generation of single-photon states as required for numerous applications in quantum communication. Tuning the fine structure splitting to zero [142] would allow to implement the method of cascaded decay of a biexciton proposed in [122] to generate polarization entangled photon pairs from an electrically pumped device. Similar to the case of recently demonstrated polarization entangled photon pair generation from optically excited QDs [123, 124], a quantum state tomography by a linear combination of cross-correlation measurements using 16 different polarization combinations may be performed with this setup. A source of entangled photon pairs on demand should allow the usage for various experiments in quantum information processing where resources like single photons and entangled photon pairs are needed. Although a similar purity of entangled state generation as in the case of parametric down-conversion has not been obtained yet, first results based on single QDs [123, 124] are very promising. As an add-on to optically excited on-demand sources, electrical excitation provides the potential for future easy-to-use as well as highly integrated devices.

Part II

Narrow-Band Single Photons
for Atomic Quantum Memories

Chapter 4

Single-Resonant OPO for Single-Photon Generation

> *"We came into the world*
> *like brother and brother;*
> *And now let's go hand in hand,*
> *not one before another."*
> – William Shakespeare

In parametric down-conversion, a pump photon incident on a nonlinear medium is split into two photons of lower frequency [40], referred to as signal and idler. Upon post-selection on the idler photons at low pump rate, this process provides a source of heralded single photons which represent the principal resource in many optical realizations of quantum information processing protocols. For quantum networks based on stationary atoms as information processing nodes and single photons to transmit information via optical

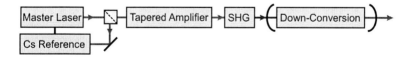

Figure 4.1: Scheme for narrow-band single-photon generation via intracavity parametric down-conversion. Emission at an atomic resonance is ensured by SHG of a master laser, stabilized to a cesium transition, and consecutive degenerate parametric down-conversion.

39

fibers, the bandwidth of down-converted photons can be reduced by mode enhancement inside an optical resonator in order to match the linewidth of atomic transitions. Additionally, the photons' central wavelength needs to coincide with the atomic resonance frequency.

In this experiment, single-photon generation on a specific atomic line in cesium (^{133}Cs) starts from second harmonic generation (SHG) of a master laser, stabilized to a ^{133}Cs transition (figure 4.1). The harmonic then acts as the pump in a degenerate down-conversion process inside a resonator. In order to achieve efficient frequency conversion and high single-photon count rates, a thorough understanding of the underlying nonlinear processes and their experimental realization is inevitable and will be subject of the following sections.

4.1   Phase-Matching

In nonlinear interactions, photons at a fundamental wavelength create photons at different frequencies, obeying energy and momentum conservation. The latter is often referred to as phase-matching. In most cases, the contributing fields can be assumed to be classical waves. Thus, the following discussion of phase-matching will be based on Maxwell theory.

The propagation of an electromagnetic wave inside a dielectric medium induces a polarization $\vec{P}$. Its components can be expressed via Taylor expansion with respect to the driving electric field $\vec{E}$

$$P_i = \epsilon_0 \left(\chi_{ij}^{(1)} E_j + \chi_{ijk}^{(2)} E_j\,E_k + \chi_{ijkl}^{(3)} E_j\,E_k\,E_l + \dots \right).$$

The coefficients $\chi^{(n)}$ represent the nth order susceptibilities and yield considerable contributions for high intensities of the fundamental mode. The second-order term involving $\chi^{(2)}$ is responsible for various important frequency mixing processes like SHG and spontaneous parametric down-conversion (SPDC). It generates secondary elementary waves which can interfere to a macroscopic field, provided proper phase relation between fundamental and harmonics is ensured. $\chi^{(2)}$ vanishes for all crystals with inversion symmetry, leaving 21 crystal classes. Not all 27 components of the second-order susceptibility are independent, though. For SHG, the indistinguishability between the two pump photons reduces their number to 18, and particular crystal symmetries can lead to further simplification.

The concept of phase-matching is most easily illustrated for the case of SHG. In order to achieve a macroscopic second harmonic excitation via constructive interference, the phase-mismatch between fundamental and harmonic

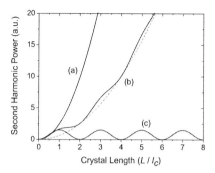

Figure 4.2: Second harmonic power inside a crystal with coherence length l_c (a) phase-matched, (b) quasi-phase-matched, and (c) unmatched.

wave

$$\Delta k = k_{2\omega} - 2k_\omega = \frac{2\omega}{c}\left(n_{2\omega} - n_\omega\right) \qquad (4.1)$$

must be minimized. For plane waves, the harmonic power for a crystal of length L follows [143]

$$P_{2\omega} \propto \mathrm{sinc}^2\left(\frac{\Delta k L}{2}\right) P_\omega^2\,.$$

A mismatch leads to different phase velocities for fundamental and harmonic waves and limits efficient conversion to the coherence length

$$l_c = \frac{\pi}{\Delta k} = \frac{\lambda}{4\left(n_{2\omega} - n_\omega\right)}\,.$$

At $l_c < L < 2l_c$, the harmonic power decreases due to reconversion into the fundamental wave which is typically of the order of a few 100 μm. Thus, special care needs to be taken to optimize phase-matching which can be adjusted via several crystal parameters. Some methods are summarized in the following paragraphs and depicted in figure 4.2.

Critical phase-matching: According to equation 4.1, plane-wave phase-matching is achieved for collinear propagation if the indices of refraction for fundamental and harmonic wave coincide ($n_\omega = n_{2\omega}$). In isotropic crystals, normal dispersion demands $n_\omega < n_{2\omega}$, and anomalous dispersion is usually limited to high-absorption materials. However, $n_\omega = n_{2\omega}$ can be achieved

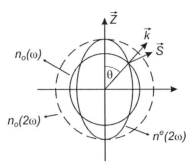

Figure 4.3: SHG phase-matching in a negative uniaxial crystal. $n_o(\omega)$ and $n^e(2\omega)$ must be equal along the propagation direction $\vec{k}$ in order to achieve efficient conversion.

in anisotropic crystals taking advantage of birefringence. The concept of birefringence is most easily understood on the basis of uniaxial crystals that show a single distinguished spatial direction, the optical axis $\vec{Z}$. A polarization perpendicular to the principal plane, spanned by $\vec{Z}$ and the wave vector $\vec{k}$ of the fundamental wave, is called ordinary while in-plane polarization is referred to as extraordinary. Thus, only the refractive index of the extraordinary wave $n^e(\theta)$ depends on the angle θ between $\vec{k}$ and $\vec{Z}$ and obeys

$$\frac{1}{[n^e(\theta)]^2} = \frac{\cos^2\theta}{n_o^2} + \frac{\sin^2\theta}{n_e^2}$$

with principal values n_o and n_e as depicted in figure 4.3 for a negative uniaxial crystal ($n_e - n_o < 0$). For phase-matching, the fundamental-wave ordinary index of refraction $n_o(\omega)$ must coincide with the harmonic-wave extraordinary index of refraction $n^e(2\omega)$.

The treatment of phase-matching in biaxial crystals is slightly more intricate since the indices of refraction along all crystallographic axes are different. This complexity, however, often simplifies to the uniaxial situation if phase-matching is solely considered in a particular principal plane [144].

If the fundamental wave in SHG or the generated waves in parametric fluorescence, respectively, consist of only ordinary or extraordinary contributions, the process is called type-I phase-matching while mixed contributions characterize a type-II process. If conversion takes place inside an optical resonator, a change of the phase-matching angle often requires a readjustment of the resonator; this led the name critical phase-matching.

In general, $\vec{k}$ is not collinear to the harmonic's Poynting vector $\vec{S}$ which is perpendicular to the tangent on the index ellipsoid at the intersection of el-lipsoid and wave vector $\vec{k}$. For ordinary polarization, $\vec{k}$ and $\vec{S}$ are parallel, but for extraordinary waves, they define an angle

$$\rho(\theta) = \pm \arctan\left(\left(\frac{n_o}{n_e}\right)\tan\theta\right) \mp \theta\,.$$

This walk-off angle leads to a divergency of different polarizations inside the crystal and limits the coherence length of nonlinear interactions. For biaxial crystals like $KNbO_3$, the walk-off also depends on the azimuthal angle. For propagation along $\vec{k}/|\vec{k}| = \hat{s} = (s_x, s_y, s_z)$, it reads

$$\rho = \arctan\left(\frac{n_{\hat{s}}^2}{\sqrt{\left(\frac{s_x}{n_{\hat{s}}^{-2}-n_x^{-2}}\right)^2 + \left(\frac{s_y}{n_{\hat{s}}^{-2}-n_y^{-2}}\right)^2 + \left(\frac{s_z}{n_{\hat{s}}^{-2}-n_z^{-2}}\right)^2}}\right)$$

with crystallographic axes x, y, and z [145].

Noncritical phase-matching: Beside crystal orientation, the crystal temperature may be used to change the indices of refraction and thus to achieve phase-matching if ordinary and extraordinary indices of refraction show different dependencies on temperature. Using temperature tuning, phase-matching can be achieved at $\theta = \pi/2$ for some materials. Then, walk-off is negligible, and the interaction region will be large. Even if phase-matching at $\theta = \pi/2$ is impossible, noncritical phase-matching is favorable for fine tuning in intra-cavity applications, as mentioned above.

Quasi-Phase-Matching: The build-up of a macroscopic polarization in unmatched crystals is limited to distances on the order of the coherence length. A renewed build-up can be initiated at multiples of the coherence length if the sign of the susceptibility switches periodically. Periodic poling has been created by a pulsed electric field, electron bombardment, thermal pulsing, or other methods [146]. The poling period for quasi-phase-matching of mth order is given by

$$\Lambda_m = \frac{2\pi}{\Delta k}\,(2m-1)\,,$$

and the corresponding nonlinearity d of this process is modified to

$$d_{eff} = \frac{2}{m\pi}\,d\,. \tag{4.2}$$

43

Since phase-matching is guaranteed via the additional momentum $2\pi/\Lambda_m$, the crystal orientation with the highest initial nonlinearity d can be chosen. Propagation of the fundamental wave along one of the optical axes also eliminates walk-off. The crystals available today are mostly fabricated by ferroelectric domain engineering; examples are lithium niobate (PPLN), lithium tantalate (PPLT), potassium titanyl phosphate (PPKTP), and potassium titanyl arsenate (PPKTA).

4.2 Nonlinear Interaction of Gaussian Beams

The description of nonlinear interactions – as presented so far – assumed plane waves. Two-photon processes like SHG are favored in the high intensity of a tight focus w_0. Besides, nearly all intra-cavity configurations place the nonlinear crystal of length L in a beam focus for symmetry reasons; thus, the theoretical treatment of the interaction requires the consideration of Gaussian beams. Characterized by the focusing parameter $\xi = L/b$ with confocal parameter b, early treatments were limited to the strong focusing $\xi \gg 1$ [147] and weak focusing $\xi \ll 1$ regime [148], respectively. The most comprehensive study of the parametric interaction between Gaussian beams has been published by Boyd and Kleinman [149]. Their work is completely applicable to the macroscopic SHG in the following experiments in order to calculate conversion efficiency κ_{NL} and harmonic power $P_{2\omega}$. However, it does not completely cover OPO processes which incorporate quantum fields. The quantum theory for an OPO below threshold will be approached separately in sections 4.5 and 5.3.

Starting from the polarization induced inside a crystal by a fundamental wave of power P_ω, the second harmonic field amplitude can be determined by cutting the crystal into infinitesimal slabs and further integrating the polarization from each slab over the crystal volume. The second harmonic power $P_{2\omega}$ is then given by

$$P_{2\omega} = \kappa_{NL}P_\omega^2 = 10^7 K L\, k_\omega \exp\left(-\alpha'L\right) h\, P_\omega^2\,. \tag{4.3}$$

Here, $\alpha' = \alpha_\omega + \alpha_{2\omega}/2$ is a weighted absorption coefficient and k_ω the wave vector of the fundamental mode. The constant

$$K = \frac{128\,\pi^2\omega^2}{(100c)^3\, n_\omega^2 n_{2\omega}} \left(\frac{3 \times 10^4}{4\pi}\, d_{eff}\right)^2$$

depends on the fundamental frequency ω, indices of refraction for the fundamental n_ω and harmonic $n_{2\omega}$ waves, and an effective nonlinear coefficient d_{eff}

which can be calculated from the susceptibility χ according to appendices 2 and 3 of [149]. The influence of the beam characteristics is described by the Boyd-Kleinman factor

$$h = \frac{1}{4\xi} \int_{-\xi(1-\mu)}^{\xi(1+\mu)} d\tau \int_{-\xi(1-\mu)}^{\xi(1+\mu)} d\tau' \frac{\exp\left(-\kappa\left(\tau + \tau'\right) + i\sigma\left(\tau - \tau'\right) - \beta^2\left(\tau - \tau'\right)^2\right)}{(1 + i\tau)(1 - i\tau')}$$

which helps to maximize the conversion efficiency. An explanation of the relevant parameters follows in the next paragraphs.

Birefringence parameter β:

$$\beta = \frac{\rho\, b}{2 w_0}$$

β describes the reduction of the conversion efficiency by walk-off, quantified by the walk-off angle ρ. For quasi-phase-matched crystals, this decrease can often be avoided.

Absorption parameter κ:

$$\kappa = \frac{b}{2}\alpha = \frac{b}{2}\left(\alpha_\omega - \frac{\alpha_{2\omega}}{2}\right)$$

Beside the factor $\exp\left(-\alpha' L\right)$ in equation 4.3, absorption must be accounted for by integrating the reduced absorption coefficient α over the crystal volume.

Focus position μ:

$$\mu = \frac{L - 2f}{L}$$

h takes its maximum if the distance f between focal position and crystal end facet is half the crystal length, i.e. for $\mu = 0$.

Phase shift σ:

$$\sigma = z_0\, \Delta k$$

In case of a residual phase-mismatch Δk between the interacting waves, the wave front curvatures of the involved beams will not evolve equally, and partial destructive interference may take place. σ characterizes the relative phase shift at the Rayleigh length z_0 of the fundamental wave.

The very same Boyd-Kleinman factor h calculates the conversion efficiency of parametric fluorescence and the threshold for OPO devices. In type-II parametric processes, however, the different indices of refraction n_S and n_I for signal and idler fields must be taken into account even at frequency-degenerate output fields which yields

$$\tilde{\kappa}_{NL} = 10^7 \tilde{K} \left(\frac{2L^2}{\tilde{w}_0^2} \right) h$$

with

$$\tilde{K} = \frac{128\,\pi^2\omega_0^2}{(100c)^3\,n_P n_S n_I} \left(\frac{3\times 10^4}{4\pi}\,2\,d_{eff} \right)^2 \quad \text{and} \quad \tilde{w}_0 = w_0 \sqrt{\frac{\omega_P}{\omega_0}} \sqrt{\frac{1}{n_S} + \frac{1}{n_I}}$$

where ω_0 and ω_P are degenerate and pump frequency, respectively. Due to the symmetries of the susceptibility tensor, $\tilde{K} = 4\,K$ holds.

Since single-photon generation from an OPO requires a very low pump rate, the threshold power is of particular interest. Equating single-pass conversion gain and corresponding losses ϵ_S, ϵ_I for signal and idler per half round-trip leads to

$$P_{thres} = \frac{\epsilon_S \epsilon_I}{\tilde{\kappa}_{NL}} .$$

The threshold power allows the calculation of OPO output rates for both classical and quantized signal and idler fields.

4.3 Optical Resonators

The output fields of frequency conversion can greatly be enhanced inside an optical resonator. This is especially interesting for two-photon processes like SHG where the harmonic power shows a parabolic dependence on the fundamental power. Different resonator geometries can be realized, each showing particular advantages. Bidirectional resonators, e.g., the Fabry-Pérot configuration, minimize the number of optical components and enhance the mechanical stability while unidirectional setups, like the bow-tie resonator, avoid reflexes back into the laser source. For maximum coupling efficiency, beam shape and input mirror reflectivity must be optimized. Length stabilization of a resonator via a feedback loop is inevitable for efficient long-term operation.

Mode-matching: Self-consistent field configurations inside a resonator are known as resonator modes and can be calculated via the resonator transfer matrix T [150, 151]. The waist of the fundamental resonator mode is derived as

$$w_0 = \sqrt{\frac{2\lambda \, |T_{1,2}|}{\pi \sqrt{4 - (T_{1,1} + T_{2,2})^2}}} \, .$$

The waist is real only for $|T_{1,1} + T_{2,2}| < 2$. In case of a Fabry-Pérot resonator of length l and mirror curvatures $R_{1/2}$, this translates into the well-known criterion

$$0 < g_1 g_2 < 1$$

for resonator stability with $g_i = 1 - l/R_i$. The shape of the incident Gaussian beam is calculated via the same matrix formalism. In order to achieve maximum coupling efficiency, the overlap between resonator mode and incident Gaussian beam must be maximized. Due to the spherical symmetry of a Gaussian beam, coupling to the resonator TEM_{00} mode is favored.

Impedance-matching: The circulating power inside the resonator

$$E_{zirk} = i\sqrt{T_{in}} E_i + g_{rt}(\omega) E_{zirk} \qquad (4.4)$$

is given by the sum of the incident field E_i transmitted through the input mirror and the circulating field of the previous round trip E_{zirk} corrected by the round-trip gain $g_{rt}(\omega)$. The round-trip gain

$$g_{rt}(\omega) = \sqrt{R_{in} R_{out}(1 - V)} \exp\left(-i\frac{\omega\, l}{c}\right)$$

depends on input (output) mirror reflectivity R_{in} (R_{out}) as well as conversion and absorption losses V. The gain is maximum on resonance, $g_{rt} = g_{rt}(\omega_{res})$, when the resonator length l equals a multiple of half the incident field wavelength $\lambda = 2\pi c/\omega$. The circulating power is then given by

$$\frac{P_{zirk}}{P_i} = \frac{T_{in}}{(1 - g_{rt})^2} \qquad (4.5)$$

with incident power P_i.

The losses $V = \bar{\alpha}l + \kappa_{NL}P_{zirk}$ are a combination of mean absorption and nonlinear conversion into harmonics or parametric fluorescence, respectively. For maximum coupling efficiency into the resonator mode, the reflected field from the input mirror

$$E_r = \sqrt{R_{in}} E_i + i\sqrt{T_{in}} \frac{g_{rt}}{\sqrt{R_{in}}} E_{zirk}$$

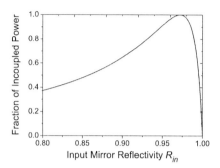

Figure 4.4: Relative power coupled into a resonator dependant on input mirror reflectivity R_{in}. This graph assumes a crystal with $3 \times 10^{-3}/\text{W}$ single-pass conversion efficiency and ~ 0.02 round-trip absorption losses at a pump power of 40 mW outside the resonator.

must be minimized which is composed of the reflected incident field E_{in} and the transmitted circulating field E_{zirk}. Using equation 4.4, the reflected field takes the form

$$\frac{E_r}{E_{in}} = \frac{R_{in} - g_{rt}}{\sqrt{R_{in}}\,(1 - g_{rt})} \, .$$

Thus, maximum coupling is achieved if the round-trip gain equals the input mirror reflectivity. This condition – together with equation 4.5 – allows the calculation of optimal impedance-matching as depicted in figure 4.4.

Hänsch-Couillaud stabilization: The most common technique to establish a lock between the length of a resonator and a laser frequency has been introduced by Pound, Drever, and Hall [152–154] which will be introduced in section 6.2. An alternative approach without the need of sideband generation is the Hänsch-Couillaud scheme [155]. However, it is only applicable to resonators that show different finesse values for orthogonal polarizations, like resonators with birefringent crystals or intra-cavity polarizers. The stabilization scheme relies on the detection of the ellipticity of a beam reflected at the resonator input mirror.

If polarizer axis and polarization of the incident field E_i span an angle α, the two components

$$E_i^{\parallel} = E_i \cos \alpha \qquad \text{and} \qquad E_i^{\perp} = E_i \sin \alpha$$

are reflected according to

$$E_r^\| = \left(\sqrt{R_{in}} - T_{in}\frac{g_{rt}(\omega)}{\sqrt{R_{in}}}\frac{1}{1 - g_{rt}(\omega)}\right)E_i^\| \qquad \text{and} \qquad E_r^\perp = \sqrt{R_{in}}\,E_i^\perp\,.$$

A linearly polarized reflected signal indicates resonance. Otherwise, the reflected signal is elliptically polarized, and the helicity of the polarization determines if the resonator length is too small or too large. In order to analyze the polarization, a quarter-wave plate (QWP) followed by a PBS is used, with the QWP fast axis set to $\pi/4$. The output modes of the PBS are monitored by a photodiode each. At resonance, i.e. with a linearly polarized reflected signal, the difference ΔI between the two detector signals yields zero. The full functional dependance of this difference can be determined using Jones calculus [156] and follows

$$\Delta I = \frac{1}{2}c\,\epsilon_0\,|E_i|^2\,2\cos\alpha\sin\alpha\frac{T_{in}g_{rt}\sin\delta}{\left(1 - g_{rt}\right)^2 + 4g_{rt}\sin^2\left(\frac{\delta}{2}\right)}$$

with round-trip phase shift $\delta = -\omega l/c$. While the maximum of this function is achieved for $\alpha = \pi/4$, the transmitting axis of the polarization selective element may be slightly shifted from optimum coupling into the resonator in order to reach both a large circulating power and a good signal-to-noise ratio of the error signal. The implementation of the feedback loop for stabilization of the resonator will be sketched in section 4.4.

4.4 Second Harmonic Generation

4.4.1 Frequency Stabilization of the Master Laser

The OPO aims at the production of frequency degenerate signal and idler photons on a hyperfine transition of the cesium D1 line at about 894.3 nm. A corresponding level scheme is provided in figure A.1. An OPO pump beam at 447.15 nm is produced by SHG, with a master laser – stabilized to the natural linewidth of a cesium resonance itself – acting as the SHG pump. The second harmonic then shows a linewidth that is $\sqrt{2}$ the fundamental value.

Laser stabilization of the master laser below the Doppler limit of cesium can be performed via the pump-probe technique of saturation spectroscopy [157]. An enhanced signal-to-noise ratio is obtained by frequency modulation spectroscopy (FMS) [158, 159] which takes advantage of lock-in detection on a phase-modulated probe beam. Figure 4.5 shows the experimental realization of the master laser stabilization. Part of the beam emitted by an

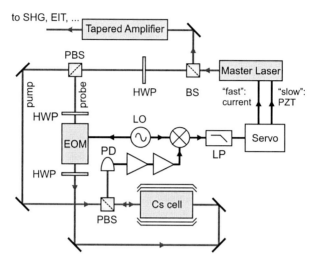

Figure 4.5: Experimental setup for sub-Doppler frequency stabilization of the OPO master laser to a cesium hyperfine transition via frequency modulation spectroscopy.

extended-cavity diode laser (DL100, Toptica Photonics) enters the stabilization setup and is split into pump and probe beams. After phase modulation by an electro-optic modulator (EOM), the probe is superimposed with the counter-propagating orthogonally polarized pump inside a 7.5 cm long cell containing 0.9999 pure ^{133}Cs. The cell windows are AR coated in order to avoid parasitic etalons. Level shift and splitting by Earth's magnetic field and stray fields, respectively, are minimized by a dual-layer μ-metal shielding that attenuates DC fields by a factor $> 10^6$. Lock-in detection of the probe modulation is performed by demodulation of the electronic signal at the local oscillator (LO) frequency, that also drives the EOM, followed by a low-pass filter. A servo, including proportional and integral filter stages, creates appropriate control signals for diode current and the extended-cavity piezo.

In order to achieve a sufficient master laser power for the subsequent experiments, 17 mW of its emission are fed into a tapered amplifier that generates 250 mW output behind a single-mode fiber to provide the pump beam for SHG.

4.4.2 KNbO₃ in a Bow-Tie Resonator

High signal and idler count rates rely on efficient SHG that is determined by the choice of nonlinear crystal and resonator geometry. First, the Boyd-Kleinman theory provides the optimal beam waist for the conversion process in a particular crystal; in a second step, a mode-matched resonator is designed for SHG pump enhancement.

The nonlinear crystal is chosen on the basis of its nonlinear coefficient. For SHG of a 894.3 nm fundamental wave, $KNbO_3$ shows the best properties in an $e + e \rightarrow o$ process. $KNbO_3$ is a biaxial crystal of point symmetry mm2 that is transparent over a wide range between 400 and 4500 nm [144, 160]. Absorption coefficients of $\alpha_\omega = 1.0$ m^{-1} and $\alpha_{2\omega} = 5.0$ m^{-1} were measured for fundamental and harmonic, respectively. Effective cw pump powers in this experiment are around 2 W, so the effects of thermal lenses and BLIIRA [161] do not need to be considered. Sellmeier equations show [144] that, in principle, noncritical phase-matching is possible between 840 and 960 nm. However, in order to minimize the requirements for temperature stabilization, the crystal is cut to allow the SHG pump to travel in its xy-plane ($\theta = 90°$) at an angle of $\phi = 35.2°$ relative to the optical axis. Phase-matching then requires a crystal temperature of 28.3 °C. The fundamental is set to extraordinary polarization while the harmonic is polarized parallel to z, i.e. at ordinary polarization. Correspondingly, the harmonic shows an elliptical shape due to walk-off of the fundamental. At the phase-matching angle, the effective nonlinear coefficient is

$$d_{eff} = d_{32} \sin^2 \left(\phi \, \frac{\pi}{180°} \right) + d_{31} \cos^2 \left(\phi \, \frac{\pi}{180°} \right) = -12.5 \, \frac{\text{pm}}{\text{V}}$$

with $d_{31} = -11.9$ pm/V and $d_{32} = -13.7$ pm/V.

As a compromise between absorption losses and harmonic power generation, a crystal length of 20 mm was chosen; its aperture measures 2.7×3.0 mm^2. From these parameters, an optimal beam waist of $w_0 = 40.0$ μm is determined with a Boyd-Kleinman factor of $h = 0.16$ resulting in a single-pass conversion efficiency of $\kappa_{NL} = 8.0 \times 10^{-3}$/W. Experiments showed a lower value, corrected by a factor of 0.42, which will be taken into account in the further considerations. According to the manufacturer, this decrease of the nonlinearity is attributed to a slight disturbance in the domain structure during the crystallization process.

The resonator is designed in bow-tie configuration. Compared to a Fabry-Pérot setup, back reflections into the laser are avoided, and the SHG power leaves the resonator into a single direction. The angles of the bow-tie are kept to a small value of $< 3.5°$ in order to minimize astigmatism. The KNbO₃

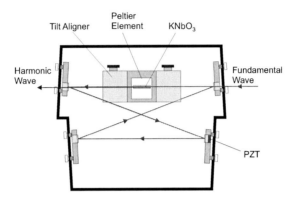

Figure 4.6: Resonator design for second harmonic generation. The cavity is resonant to the fundamental wave while the harmonic wave leaves the resonator instantly.

crystal is placed in the focus between two curved mirrors with 80 mm curvature which define one resonator arm of length 0.103 m. The second arm, including the two plane mirrors, measures 0.365 m. Using the ray transfer matrix method, a self-consistent resonator mode with the desired waist $w_0 = 40$ μm between the curved mirrors is derived. The total optical length of the ring resonator is $L_{opt} = 493$ mm, corresponding to an FSR = 608 MHz. In order to achieve a high circulating pump power P_{zirk}, the resonator must be impedance-matched via the input coupler reflectivity $R_{1\omega}$. The round-trip gain

$$g_{rt} = \sqrt{R_{1\omega} R_{2\omega} R_{3\omega} R_{4\omega} \left(1 - \kappa_{NL} P_{zirk} - \alpha_\omega L - 2 R_{AR}\right)}$$

take into account mirror reflectivities $R_{i\omega} > 0.999$ ($i = 2, 3, 4$), conversion and absorption losses, and imperfect anti-reflection coatings $R_{AR} \approx 10^{-3}$ on the crystal facets. Together with $R_{1\omega} \equiv g_{rt}(\omega)$, an optimal reflectivity $R_{1\omega} = 0.967$ and circulating power $P_{zirk} = 2.49$ W are calculated for 100 mW pump power. The measured reflectivity $R_{1\omega} = 0.985$ and a non-unity coupling efficiency into the resonator of 0.81 (see subsection 4.4.3) allows a circulating power $P_{zirk} = 2.22$ W only, though.

The resonator design needs to consider temperature stability and isolation from acoustic noise (figure 4.6). Thus, the external case is manufactured from a single aluminum block; a plexiglass lid and AR coated windows for optical access limit air turbulence. The mirror mounts are directly attached to the case walls, one of the plane mirrors is glued to two stacked PZT elements that allow the scan over one FSR with a peak-peak voltage of about 30 V.

A four-axes tilt aligner sets the phase-matching angle of the $KNbO_3$ crystal that is held inside a copper block, temperature stabilized within ± 25 mK.

4.4.3 SHG Characterization

The generation of error signals for cavity stabilization and pump power enhancement requires an optimum coupling between pump beam and resonator mode. By periodically scanning the PZT mounted to one of the cavity mirrors, the reflected signal in figure 4.7(a) was measured at a pump power of 45 mW. From the FSR value of 608 MHz, a linewidth of $\delta\nu = 4.5$ MHz and a corresponding finesse of 135 can be derived. This is close to the expected finesse of 145 for the measured input mirror reflectivity $R_{1\omega} = 0.985$; the deviation may be caused by crystal absorption that exceeds the reported value [144]. The reflection signal shows that higher transversal modes of the resonator are efficiently suppressed. The coupling efficiency to the TEM_{00} mode is 0.81 which perfectly agrees with the theoretical maximum of 0.814 at 45 mW pump power due to impedance matching with $R_{1\omega} = 0.985$ (contribution: 0.855) and the convolution of a 300 kHz laser linewidth with a 4.5 MHz cavity resonance (contribution: 0.953).

By detecting the polarization ellipticity of the beam reflected from the input mirror, an error signal for resonator stabilization is created according to section 4.3. At the same pump power of 45 mW, the PZT was scanned over one FSR yielding the dependance depicted in figure 4.7(b). The signal-to-noise ratio exceeds 100 : 1, and cavity stabilization is possible over several hours, only limited by thermal drifts that require voltages beyond the available PZT range.

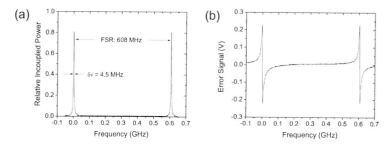

Figure 4.7: (a) Relative incoupled power at a scan of the SHG resonator length over one FSR. (b) Corresponding error signal used for resonator stabilization.

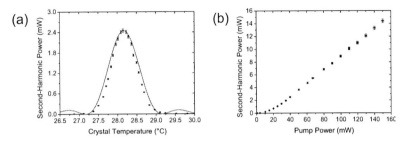

Figure 4.8: Second harmonic power dependance on (a) KNbO$_3$ crystal temperature and (b) incident fundamental power.

Long-term temperature stabilization of the KNbO$_3$ crystal is crucial for phase-matching and continuous SHG operation. Thus, the dependance of SHG power on crystal temperature was measured at 45 mW pump power as depicted in figure 4.8(a). The theoretical curve relies on Sellmeier equations [144] which fit the measurement assuming a phase-matching angle $\phi = 35.22°$. The experimental data are well reproduced with a theoretical temperature acceptance range of 0.9 K slightly larger than the measured value of 0.75 K. Thus, temperature fluctuations, which are controlled within ± 25 mK inside the KNbO$_3$ crystal, result in second harmonic power deviations of < 0.01.

The dependance of second harmonic power on pump power is provided in figure 4.8(b). At low pump levels, the output power follows a parabolic shape according to single-pass theory, while increasing conversion losses lead to a linear evolution at higher pump rates. The measurement was taken at a crystal temperature of 28.22 °C, and up to 14 mW of blue light are extracted behind the SHG setup. In order to pump the OPO, a mode-cleaning fiber behind the SHG setup is inevitable to remove higher transversal mode contributions due to walk-off inside the KNbO$_3$ crystal. Up to 7 mW SHG emission in a TEM$_{00}$ could be obtained which already exceeds the SRO pump rate for the "far below threshold" regime which will be subject of the next section.

4.5 Single-Photon Generation from an SRO

The generation of the OPO pump as well as the optimization of mode-matching of the pump to the OPO resonator can be treated classically. This section provides the theoretical background to understand the generation of the non-classical signal-idler biphoton state. From this state, observables

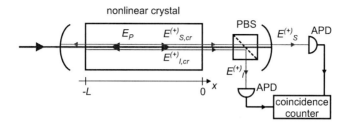

Figure 4.9: Schematic setup of the SRO. The cavity acts as a one-sided resonator for the signal field while the idler field is non-resonant due to deflection at an intra-cavity PBS. Signal-idler coincidences can be determined by an HBT correlator.

like count rate, spectra of the parametric fluorescence, and the second-order signal-idler cross-correlation function can be derived by applying the standard perturbative treatment in the Schrödinger picture [40]. Different from former approaches used for the theoretical description of the double-resonant OPO [43, 162, 163], free-field quantization of the idler field is necessary for a single-resonant OPO (SRO) far below threshold.

The experimental situation is a type-II parametric down-conversion process in a nonlinear crystal of length L, pumped by a linearly polarized laser with frequency ω_P. The surrounding cavity is resonant for the signal field, but not for the idler field which is polarized orthogonally to the signal. Figure 4.9 shows a schematic of the corresponding experimental setup.

The interaction Hamiltonian: As introduced in section 4.1, energy and momentum must be conserved for a conversion process which reads

$$\begin{aligned} \omega_P &= \omega_S + \omega_I, \\ \vec{k}_p(\omega_P) &= \vec{k}_s(\omega_S) + \vec{k}_i(\omega_I). \end{aligned} \tag{4.6}$$

with central frequencies ω_S and ω_I for signal and idler field, respectively [40]. The wave vectors $\vec{k}_p$, $\vec{k}_s$, and $\vec{k}_i$ of pump, signal, and idler wave are considered collinear. The idler field leaves the cavity in positive x-direction while the standing-wave fields of pump and signal inside the cavity consist of two counter-propagating components. Due to equation 4.6, only the components traveling collinear to the idler wave need to be taken into account. The positive-frequency part of the classical pump field inside the crystal can be written as

$$E_{P,cr}(x,t) = E_P \, e^{i[k_p(\omega_P)\,x - \omega_P t]}.$$

On the other hand, the positive-frequency signal and idler fields inside the crystal need a quantum mechanical treatment and are described by operators $E_{S,cr}^{(+)} = E_{S,cr}^{(-)\dagger}$ and $E_{I,cr}^{(+)} = E_{I,cr}^{(-)\dagger}$, respectively. This notation leads to the interaction Hamiltonian

$$H_{int} = \frac{\chi}{2l} \int_{-L}^{0} dx \left(E_{P,cr} E_{S,cr}^{(-)} E_{I,cr}^{(-)} + E_{P,cr}^{*} E_{S,cr}^{(+)} E_{I,cr}^{(+)} \right) \qquad (4.7)$$

where the second-order susceptibility χ is frequency dependent [40]. Further evaluation of equation 4.7 will require explicit expressions for the operators $E_{S,cr}^{(+)}$ and $E_{I,cr}^{(+)}$.

The free-field operators: If A is the transverse cross-section of a wave propagating in x-direction, its electric field operator can be written as

$$E^{(+)}(x,t) = \lim_{L \to \infty} \sum_{j=0}^{\infty} \sqrt{\frac{\hbar \omega_j}{2\epsilon_0 L A}} \, a_j \, e^{i\omega_j \left(\frac{x}{c} - t \right)}$$

where $\omega_j = 2\pi j c/L$ with quantization length L and a_j denotes the photon annihilation operator of mode j. The limit can be expressed by the replacement

$$\sum_j a_j \ldots \to (\Delta\Omega)^{-1/2} \int_{-\omega_S}^{\infty} d\Omega \, a(\omega_S + \Omega) \ldots$$

with $\Omega_j = \omega_j - \omega_S$ and corresponding $\Delta\Omega = 2\pi c/L$. The typical down-conversion bandwidth is small compared to the signal and idler central frequencies which justifies an integration interval extended to $-\infty$:

$$E_S^{(+)}(x,t) = \sqrt{\frac{\hbar \omega_S}{2\epsilon_0 c A}} \int_{-\infty}^{\infty} \frac{d\Omega}{\sqrt{2\pi}} \, a(\omega_S + \Omega) \, e^{i(\omega_S + \Omega)\left(\frac{x}{c} - t \right)}$$

where [75]

$$[a(\omega_1), a^\dagger(\omega_2)] = \delta(\omega_1 - \omega_2).$$

An equivalent expression describes the idler field in free space

$$E_I^{(+)}(x,t) = \sqrt{\frac{\hbar \omega_I}{2\epsilon_0 c A}} \int_{-\infty}^{\infty} \frac{d\Omega}{\sqrt{2\pi}} \, b(\omega_I + \Omega) e^{i(\omega_I + \Omega)\left(\frac{x}{c} - t \right)}$$

with

$$[b(\omega_1), b^\dagger(\omega_2)] = \delta(\omega_1 - \omega_2)$$

where a and b are mutually orthogonal due to the type-II nature of the considered process.

The field operators inside the crystal: If a lossless dispersive medium
is present [40], the idler field inside the crystal takes the form [164]

$$E^{(+)}_{I,cr}(x,t) = \sqrt{\frac{\hbar\omega_I}{2\epsilon_0 cAn_I}} \int_{-\infty}^{\infty} \frac{d\Omega}{\sqrt{2\pi}}\, b(\omega_I + \Omega)\, e^{i[k_I(\Omega)x - (\omega_I + \Omega)t]} \quad (4.8)$$

with

$$k_I(\Omega) = \frac{\omega_I + \Omega}{c}\, n_i(\omega_I + \Omega)\,.$$

The influence of the medium is taken into account by $\epsilon_0 \to \epsilon_0 n_i^2$ and $c \to c/n_i$
with the refractive index $n_i(\omega)$ of the idler wave and $n_I = n_i(\omega_I)$.
If the resonator is assumed to be completely filled by the nonlinear medium
of length L, the expression for the signal field with index of refraction $n_s(\omega)$
reads [165]

$$E^{(+)}_{S,cr}(x,t) = \sqrt{\frac{\hbar\omega_S \Delta\omega_c}{\epsilon_0 n_S cA\pi}} \sum_{m=-\infty}^{\infty} a_m \frac{e^{i\omega_m\left[\frac{x}{c}n_s(\omega_m)-t\right]}}{2}$$

with resonator frequencies ($m \in \mathbb{Z}$, $m_0 \gg |m|$)

$$\omega_m \approx \omega_S + m\,\frac{\pi\, v_{g,S}}{L} \equiv \omega_S + m\,\Delta\omega_c\,.$$

$v_{g,S}$ is the group velocity of the signal field at frequency ω_S

$$v_{g,S} = \frac{c}{n_S + \omega_S \frac{\partial n_s}{\partial\omega}|_{\omega=\omega_S}}$$

and $n_S = n_s(\omega_S)$.   The photon annihilation and creation operators for
mode m obey the usual commutation relation $[a_m, a^\dagger_{m'}] = \delta_{m,m'}$.
Standard input-output formalism [75, 166] can be applied if resonator losses
of rate γ are incorporated. The now time-dependant modes read

$$a_m(t) = \frac{1}{\sqrt{2\pi}} \int_{-\infty}^{\infty} d\Omega\, a(\omega_m + \Omega)\, \frac{\sqrt{\gamma}}{\frac{\gamma}{2} + i\Omega}\, e^{-i\Omega t}$$

with their commutator in Fourier space

$$[a(\omega_m + \Omega), a^\dagger(\omega_{m'} + \Omega')] = \delta_{m,m'}\delta(\Omega - \Omega')\,.$$

This commutator implies that the quasi-modes do not overlap spectrally
which is justified in the good-cavity limit

$$\gamma \ll \Delta\omega_c\,.$$

57

This finally yields the signal operator inside the lossy cavity

$$E_{S,cr}^{(+)}(x,t) \;=\; \sqrt{\frac{\hbar\omega_S}{2\epsilon_0 n_S c A}}\,\frac{\sqrt{\gamma\Delta\omega_c}}{2\pi}\,\times \tag{4.9}$$

$$\times\;\sum_{m=-\infty}^{\infty}\int_{-\infty}^{\infty}d\Omega\,\frac{a(\omega_m+\Omega)}{\frac{\gamma}{2}+i\Omega}\,e^{i\left[k_{S,m}(\Omega)x-(\omega_m+\Omega)t\right]}$$

with the wave vector of frequency $\omega_m+\Omega$

$$k_{S,m}(\Omega) \;=\; \frac{\omega_m+\Omega}{c}\,n_s(\omega_m+\Omega)\,.$$

The denominator of equation 4.9 indicates resonance enhancement at $\omega\approx\omega_m$. For a resonator longer than the crystal, a rigorous quantization of the signal field has to account for the exact position of the crystal inside the resonator. However, it is sufficient to describe the signal field inside the crystal by equation 4.9 with $\Delta\omega_c=\Delta\omega$ and $\omega_m=\omega_S+m\Delta\omega$ where

$$\Delta\omega=\frac{2\pi}{T}\qquad\text{with}\qquad T=\frac{2l}{v_{g,S}}+\frac{2\left(L_r-l\right)}{c}$$

is the effective free spectral range.

Explicit form of the Hamiltonian: Equations 4.8 and 4.9 for idler and signal field operators can now be substituted into equation 4.7 to obtain the explicit form of the Hamiltonian

$$H_{int} \;=\; i\hbar\alpha\sum_{m=-\infty}^{\infty}\int_{-\infty}^{\infty}d\Omega\,\frac{\sqrt{\gamma}}{\frac{\gamma}{2}-i\Omega}\int_{-\infty}^{\infty}d\Omega'\,F_m(\Omega,\Omega')\,\times$$

$$\times\;a^\dagger(\omega_m+\Omega)b^\dagger(\omega_I+\Omega')\,e^{i(m\Delta\omega+\Omega+\Omega')t}\;+\;\text{h.a.}$$

with definitions

$$F_m(\Omega,\Omega') \;=\; \frac{\chi(\omega_P;\omega_m+\Omega,\omega_I+\Omega')}{\chi(\omega_P;\omega_S,\omega_I)\,l}\int_{-l}^{0}dx\,e^{i\left[k_P-k_{S,m}(\Omega)-k_I(\Omega')\right]}$$

$$\alpha \;=\; \frac{-iE_P}{8\pi\epsilon_0 c A}\sqrt{\frac{\omega_S\omega_I}{n_S n_I}}\,\chi(\omega_P;\omega_S,\omega_I)\sqrt{\Delta\omega}\,.$$

In the low-pump regime with photon generation rate $\kappa\ll\gamma$, the formal solution of the Schrödinger equation can be expanded [40, 167, 168]

$$|\psi(\Delta t)\rangle \;\propto\; |0\rangle+|\tilde{\psi}(\Delta t)\rangle=|0\rangle+\frac{1}{i\hbar}\int_0^{\Delta t}dt\,H_{int}(t)|0\rangle\,,$$

where $|\tilde{\psi}(\Delta t)\rangle$ represents the perturbation of the vacuum $|0\rangle$ due to biphoton generation, and the integral evaluates to

$$
\begin{aligned}
|\tilde{\psi}(\Delta t)\rangle \;=\; & \sum_{m=-\infty}^{\infty} \int_{-\infty}^{\infty} d\Omega \, \frac{\alpha\sqrt{\gamma}}{\frac{\gamma}{2} - i\Omega} \int_{-\infty}^{\infty} d\Omega' F_m(\Omega, \Omega') \\
& \times \Delta t \, \text{sinc}\left[\frac{1}{2}(m\Delta\omega + \Omega + \Omega')\Delta t\right] e^{\frac{i}{2}(m\Delta\omega + \Omega + \Omega')\Delta t} \\
& \times a^\dagger(\omega_m + \Omega) b^\dagger(\omega_I + \Omega')|0\rangle .
\end{aligned}
\tag{4.10}
$$

Equation 4.10 corresponds to the state of the radiation field on the condition that a signal-idler photon pair has been produced during the time interval Δt. The sinc-function in equation 4.10 may be replaced by $2\pi\,\delta(m\Delta\omega + \Omega + \Omega')$ according to [168]. One then obtains the wave function [165]

$$
\begin{aligned}
|\psi\rangle \;=\; & \mathcal{N} \sum_{m=-\infty}^{\infty} \int_{-\infty}^{\infty} d\Omega \, \frac{\Phi_m(\Omega)}{\frac{\gamma}{2} - i\Omega} \\
& \times a^\dagger(\omega_S + m\Delta\omega + \Omega) \, b^\dagger(\omega_I - m\Delta\Omega - \Omega)|0\rangle
\end{aligned}
$$

where the explicit value of $\mathcal{N}$ is not important as long as only normalized quantities characterizing the radiation field are considered. The phase-matching function $\Phi_m(\Omega)$ is denoted by

$$
\Phi_m(\Omega) \;=\; F_m(\Omega, -m\Delta\omega - \Omega) \approx \frac{1}{l} \int_{-l}^{0} dx \, e^{i(m\Delta\omega + \Omega)\frac{\tau_0}{l}x}. \tag{4.11}
$$

with the difference

$$
\tau_0 \;=\; \frac{l}{c}\left(n_I + \omega_I \frac{\partial n_i}{\partial\omega}\bigg|_{\omega=\omega_I} - n_S - \omega_S \frac{\partial n_s}{\partial\omega}\bigg|_{\omega=\omega_S} \right) \tag{4.12}
$$

between the transit times of signal and idler photons through the crystal.

Photon generation rate: The approximation in equation 4.11 holds in the limit $\gamma \ll \Delta\omega \ll \tau_0^{-1}$. Due to $\gamma \ll \Delta\omega$, equation 4.11 can further be approximated by

$$
\Phi_m(\Omega) \approx \Phi_m(0) = \text{sinc}\left(m\Delta\omega\frac{\tau_0}{2}\right) e^{-im\Delta\omega\frac{\tau_0}{2}} .
$$

Now, the photon generation rate follows as [165]

$$
\begin{aligned}
\kappa \;=\; & \frac{1}{\Delta t} \langle \tilde{\psi}(\Delta t)|\tilde{\psi}(\Delta t)\rangle \\
=\; & \left(\frac{\chi|E_p|}{4\epsilon_0 cA}\right)^2 \frac{\omega_S\omega_I}{n_S n_I} \Delta\omega \sum_{m=-\infty}^{\infty} \text{sinc}^2\left(m\Delta\omega\frac{\tau_0}{2}\right) \\
\approx\; & \left(\frac{\chi|E_p|}{4\epsilon_0 cA}\right)^2 \frac{\omega_S\omega_I}{n_S n_I} \frac{2\pi}{|\tau_0|} .
\end{aligned}
\tag{4.13}
$$

Spectral properties: The spectra of signal and idler fields are defined as

$$S_{S/I}(\omega) = \frac{1}{2\pi} \int_{-\infty}^{\infty} d\tau \, G_{S/I}^{(1)}(\tau) \, e^{i\omega\tau}$$

where the first-order correlation function can be derived from the biphoton wave function and the free-field signal and idler operators

$$G_{S/I}^{(1)}(\tau) = \langle\psi| E_{S/I}^{(-)}(x,t) E_{S/I}^{(+)}(x,t+\tau) |\psi\rangle.$$

For the idler field, this yields

$$G_I^{(1)}(\tau) \propto \sum_{m=-\infty}^{\infty} \int_{-\infty}^{\infty} d\Omega \, \frac{|\Phi_m(\Omega)|^2}{\left(\frac{\gamma}{2}\right)^2 + \Omega^2} e^{-i(\omega_I - m\Delta\omega - \Omega)\tau}$$

and after Fourier transform

$$S_{S/I}(\omega) \propto \sum_{m=-\infty}^{\infty} \frac{\mathrm{sinc}^2\left(m\Delta\omega\frac{\tau_0}{2}\right)}{\left(\frac{\gamma}{2}\right)^2 + (\omega_{S/I} - m\Delta\omega - \omega)^2}. \qquad (4.14)$$

An illustration of this spectral dependance is depicted in figure 4.10. Its overall bandwidth is determined by the phase-matching condition of width $\approx \tau_0^{-1}$. Both signal and idler spectra are composed of Lorentzians of FWHM γ centered at frequencies $\omega_{S/I} - m\Delta\omega$ with $m \in \mathbb{Z}$. Only the signal is resonant; however, the longitudinal mode structure of the resonator translates into the idler spectrum due to the frequency correlation between signal and idler fields.

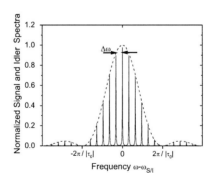

Figure 4.10: Schematic plot of the normalized signal and idler output spectra. The width of the Lorentzian peaks is determined by the cavity damping rate γ, and they are separated by the free spectral range (FSR) $\Delta\omega = 2\pi/T$. The envelope, described by the sinc-function in equation 4.14, yields the total spectral width $2\pi/|\tau_0|$.

Signal-idler intensity cross-correlation: The temporal second-order cross-correlation function

$$G_{IS}^{(2)}(\tau) = \langle\psi|E_I^{(-)}(x,t)E_S^{(-)}(x,t+\tau)\times$$
$$\times E_S^{(+)}(x,t+\tau)E_I^{(+)}(x,t)|\psi\rangle \qquad (4.15)$$

can be calculated in an equivalent way using the free-field operators of both signal and idler fields. Using residue integration, the following expression is derived [165]

$$G_{IS}^{(2)}(\tau) \propto \begin{cases} e^{-\gamma\tau}\left|\sum_m \mathrm{sinc}\left(m\Delta\omega\frac{\tau_0}{2}\right)e^{-im\Delta\omega\left(\tau+\frac{\tau_0}{2}\right)}\right|^2 & \text{if } \tau + \frac{\tau_0}{2} \geq -\frac{|\tau_0|}{2} \\ 0 & \text{if } \tau + \frac{\tau_0}{2} < -\frac{|\tau_0|}{2} \end{cases} \qquad (4.16)$$

This function consists of a train of exponentially decaying peaks spaced by the resonator round-trip time $T = 2\pi/\Delta\omega$. The individual peaks have a width of τ_0. The time shift $\tau_0/2$ of the peaks arises since for $\tau_0 > 0$ the time needed by the center of the signal wave packet to travel from the middle of the crystal to its end facet is by the amount $\tau_0/2$ shorter than the time needed by the center of the idler wave packet to cover the same distance while the opposite holds for $\tau_0 < 0$.
Transforming the sum in equation 4.16 into an integral yields the simplified

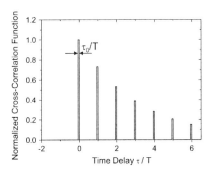

Figure 4.11: Schematic of the normalized second-order signal-idler cross-correlation function $G_{IS}^{(2)}(\tau)$ according to equation 4.17 for $\tau_0 > 0$ and $\gamma/\Delta\omega = 0.05$. If the time delay τ is equal to zero or multiples of the cavity round-trip time $T = 2\pi/\Delta\omega$, the function exhibits pronounced peaks which decay with the cavity damping time γ^{-1}.

form

$$G_{IS}^{(2)}(\tau) \quad \propto \quad \sum_{j=0}^{\infty} \begin{cases} e^{-\gamma j T} & \text{if } \left| \tau - j\,T + \frac{\tau_0}{2} \right| \leq \frac{|\tau_0|}{2} \\ 0 & \text{else} \end{cases} \qquad (4.17)$$

which displays the physical content of the correlation function more obviously as shown in figure 4.11. The comparison with experimental data will need to take into account the limited time resolution of the HBT setup by convoluting equation 4.17 with a Gaussian distribution.

4.6 Setup of the Single-Resonant OPO

4.6.1 Design Considerations

The main design parameters of an OPO pertain to the nonlinear crystal and resonator geometry. Spatial separation of signal and idler fields on the cesium D1 line requires a crystal that allows a degenerate type-II process at 894.3 nm. For this experiment, monoclinic bismuth triborate (BiB_3O_6, BiBO) was chosen which shows a wide transmission range between 286 and 2500 nm, high damage threshold, no hygroscopicity, and a high nonlinear coefficient at the desired wavelength. Phase-matching can be achieved at room temperature in the xz-plane via an $o \rightarrow e + o$ process with ordinary polarization for pump and signal waves. The exact phase-matching angle ϕ inside this plane ($\theta = 0$) is calculated from Sellmeier equations [169] and depicted in figure 4.12(a); it reaches degenerate emission for $\phi = 65.32°$. The walk-off at this angle is specified to 70.5 mrad, and an effective nonlinear coefficient $d_{eff} = 1.45$ pm/V is determined for degenerate output frequencies [170]. Due to its wide transmission range, BiBO is well suited for nonlinear processes. Unfortunately, wavelength-dependant absorption coefficients are not available and only a rough estimate of $< 10/$m at 900 nm and $\approx 10/$m at 450 nm could be provided by the manufacturer[1].

The BiBO crystal length was chosen as 20 mm with an aperture of 3.0×2.8 mm^2 with AR coated facets which limit residual reflections to $< 2 \times 10^{-3}$ at the wavelengths of interest. From Boyd-Kleinman theory, an optimum pump waist $w_0 = 33$ μm for these crystal parameters can be derived as depicted in figure 4.13(a). The non-unity value of the Boyd-Kleinman factor is mainly due to birefringence and wave-front mismatch between pump and output waves.

Regarding resonator design, a Fabry-Pérot configuration was favored for the

---

[1] FEE GmbH

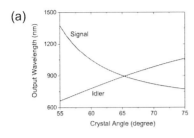 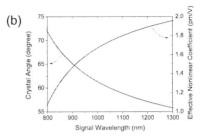

Figure 4.12: (a) Signal and idler wavelengths dependance on crystal angle ϕ at critical phase-matching. Degenerate output is achieved for $\phi = 65.32°$. (b) Variation of the effective nonlinear coefficient d_{eff} with the angle-dependant signal output wavelength. At degenerate output (894.3 nm), $d_{eff} = 1.45$ pm/V can be achieved.

SRO cavity. Here, the unidirectional propagation of a bow-tie resonator is less important since the circulating pump power is smaller than in the SHG setup and back scattering plays a minor role. A Fabry-Pérot cavity, on the contrary, allows a more compact design with higher mechanical stability and resistance to temperature drifts. In order to achieve maximum output power per resonator mode, a large FSR is favored with a correspondingly small effective resonator length. On the other hand, a narrow mode bandwidth requires a small FSR.

The resonator needs to accommodate the BiBO crystal as well as a PBS in order to deflect the idler out of the resonator mode; thus, a mirror distance of at least 60 mm is required. For this setup, mirror curvatures of 80 mm offer the highest versatility with stable resonator lengths between 10 and 170 mm and a range insensitive to astigmatism around 90 mm (figure 4.13(b)). To account for the higher Boyd-Kleinman factor at smaller pump waists, however, a mirror distance of 0.11 m was chosen, corresponding to an effective resonator length for the signal of 131 mm. With this length, stable resonance conditions for both the pump and the signal are achieved.

Before the resonator linewidth can be determined, impedance-matching must be optimized via the input mirror reflectivity. The round-trip gain g_{rt} for the pump beam consists of absorption α_p inside the BiBO crystal, residual AR coating losses ($< 2 \times 10^{-3}$ per facet), and limited transmission of the PBS (≈ 0.98). This combination yields $g_{rt} = 0.76$. The transmission of the output mirror and conversion losses can be neglected for the typical pump beam power of 1 mW. The signal output mirror reflectivity was chosen to 0.985 as a trade-off between high signal count rates and a narrow linewidth.

63

(a)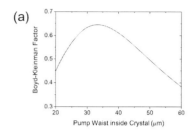
(b)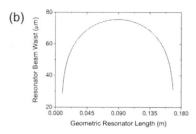

Figure 4.13: (a) Dependance of the Boyd-Kleinman factor on beam waist inside the nonlinear crystal. The optimal waist is $w_0 = 32$ μm. (b) Beam waist for a stable resonator configuration at different resonator lengths. A waist at the upper plateau around 75 μm is favored to minimize astigmatism. Eventually, an experimental beam waist $w_0 = 69$ μm was chosen.

From resonator parameters, the pump beam finesse is calculated to $F_P = 10$; an incident beam of 1 mW thus leads to a circulating power of 2.3 mW. A finesse $F_S = 21$ can be determined for the signal, resulting in an expected signal linewidth of 53 MHz.

4.6.2 Experimental Setup and Resonator Characterization

The experimental design follows the scheme presented in figure 4.9. The SRO is set up inside an air-tight box in order to protect against acoustic noise; a top view is provided in figure 4.14. The ground plate is temperature stabilized within ± 25 mK, and small optical components with high resonance frequencies were chosen to minimize mechanical vibrations. BiBO crystal and PBS are mounted on top of a five-axes kinematic stage to adjust the phase-matching angle. The nonlinear crystal is separately temperature stabilized. The input mirror is glued onto a PZT for length stabilization of the mirror distance. The pump power reflected from the input mirror is depicted in figure 4.15(a) when the PZT is linearly scanned over several FSR. A coupling efficiency of 0.28 was measured; this low value can be explained by the input mirror reflectivity of 0.935 which was ordered at a time when lower pump absorption losses in BiBO were expected from preliminary manufacturer specifications. With an adequate reflectivity, that matches the pump round-trip gain $g_{rt} = 0.76$, a value comparable to the SHG coupling efficiency can be reached, but the resulting circulating power is still sufficient

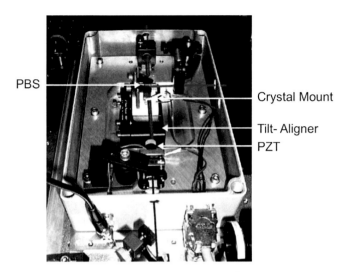

Figure 4.14: Photograph of the SRO setup. For better visualization, pump (blue), signal (red), and idler (green) beam paths are sketched.

for cavity length stabilization. A resonator finesse $F_P = 9.8$ for the pump beam is derived from the reflected beam which agrees well with the calculated value. The pump beam reflected from the input mirror is also used for resonator length stabilization via the Hänsch-Couillaud scheme as described in section 4.3.

Figure 4.15(b) shows typical error signals. The signal-to-noise ratio of $10 : 1$ can be attributed to the low pump power of the SRO, the small 0.06 fraction of the reflected pump power that enters the stabilization setup behind a beam sampler, and the inferior detection efficiency of FND-100 photo diodes in the blue. However, these error signals allowed the stabilization of the SRO resonator over several hours.

4.6.3 Count Rate and Correlation Measurements

For characterization of the SRO emission, signal and idler fields are separated from the pump beam by long-pass filters as well as additional 1 nm and 10 nm wide band pass filters to absorb spurious near-infrared fluorescence from the long-pass filters. Signal and idler fields are detected by APD modules for subsequent temporal correlation measurements. An idler detection

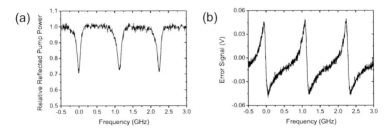

Figure 4.15: (a) Relative reflected pump power from the SRO input mirror. The absorption signals reveal a pump beam finesse of $F_P = 9.8$. (b) Error signal for resonator stabilization generated via the Hänsch-Couillaud scheme.

event triggers the start of the coincidence counter[2] described in section 2.3; a signal photon, which arrives simultaneously or after a number of round trips inside the resonator, stops the counter. Similar to auto-correlation measurements, a histogram dependant on idler-signal delays is obtained with 40 ps time bins.

Measurements are depicted in figure 4.16 which provides information about principal parameters like resonator decay time, given by the inverse of the resonator bandwidth, and its free spectral range. In order to fit equation 4.17 to the experimental data, a coincidence offset caused by coherent background light as well as a time delay offset taking into account different electronic path lengths in the correlation setup have been added. Furthermore, the time resolution of the correlation setup must be incorporated. Thus, the solid line in figure 4.16(a) shows a convolution of the theoretical $g_{IS}^{(2)}$ function with a Gaussian of width $t_{res} = 800$ ps. This time resolution was determined by a previous auto-correlation measurement of fs laser pulses. A resonator bandwidth $\gamma \approx 62$ MHz and a free spectral range $\Delta\omega \approx 1.15$ GHz (corresponding to the round-trip time $T_{00} = 867$ ps) can be derived from this fit with excellent agreement between measurement data and theoretical predictions. From our data, exact values for the BiBO absorption coefficient can be found at 450 nm ($\alpha_S = 10.2$ m^{-1}) and 894 nm ($\alpha_S = 6.0$ m^{-1}).

Figure 4.16(b) provides a second data set that shows a spacing between subsequent maxima of $T_{10} \approx 1750$ ps which is twice the resonator round-trip time. Here, the output beam was spatially filtered by a pinhole as shown in the inset. This behavior can be interpreted by the excitation of a higher transversal resonator mode, e.g., the TEM$_{10}$ mode, which hits the pinhole

---

[2]TimeHarp 200, PicoQuant

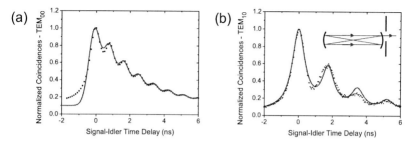

Figure 4.16: (a) Second-order cross-correlation function between signal and idler photons for the TEM_{00} mode. The periodic structure traces the cavity round-trip time. A cavity bandwidth of 62 MHz can be derived by the exponential decay of the envelope. The solid line is obtained by a theoretical model of the correlation function as described in the text. (b) Experimental results of the second-order cross-correlation function with pinhole filtering of the TEM_{10} mode; the inset depicts the position of the pinhole. A doubled round-trip time can be observed.

only at every second round-trip due to its spatial distribution. Again, a resonator bandwidth $\gamma \approx 62$ MHz fits the experimental data.

The correlation measurements were taken at a pump power of 1 mW and an experimental coincidence count rate of

$$\kappa_{coin}^{exp} = 660 \text{ counts}/(\text{s mW})$$

could be obtained. To explain this value by the theoretical predictions of equation 4.13, the Boyd-Kleinman factor h and setup attenuation need to be taken into account. The pump beam waist inside the crystal was set to 69 μm, corresponding to the optimum value for a resonator length of 131 mm. Considering this waist and the birefringence parameter of BiBO at the phase-matching angle, $h = 0.36$ is calculated. The setup transmission for both signal and idler $T_{S/I}$ is the product of APD detection efficiency $\eta = 0.35$, transmissions of long-pass ($T_{LP} = 0.83$) and band pass ($T_{BP} = 0.75 \times 0.63$) filters, as well as resonator escape efficiencies for signal and idler, respectively. With a nonlinear susceptibility $\chi = 2\epsilon_0 d_{eff} V$, where V is the resonator mode volume, a theoretical coincidence rate

$$\kappa_{coin}^{theo} = h \, \kappa \, T_S T_I = 694 \text{ counts}/(\text{s mW})$$

is calculated according to equation 4.13. This value fits the experimental rate κ_{coin}^{exp} extremely well, especially since the theoretical prediction is obtained from first-principle calculations. Due to the high absorption losses in BiBO and the correspondingly differing escape efficiencies for signal and idler, count rates behind the resonator output mirror of

$$\kappa_S^{exp} = 6.7 \times 10^4 \text{ counts/s} \quad \text{and} \quad \kappa_I^{exp} = 2.9 \times 10^5 \text{ counts/s}$$

have been measured at 1.5 mW pump power for signal and idler, respectively, which is reproduced by theoretical predictions of

$$\kappa_S^{th} = 5.87 \times 10^4 \text{ counts/s} \quad \text{and} \quad \kappa_I^{th} = 2.88 \times 10^5 \text{ counts/s}.$$

These values lead to a pair production generation rate $\kappa' = \kappa \times h = 2.04 \times 10^5$ counts/(s mW) inside the resonator.

4.6.4 Discussion

The presented SRO far below threshold has been operated as a compact source of narrow-band single photons. Its advantage over previous realizations of cavity-enhanced parametric down-conversion is its continuous and reliable stabilization of the SRO resonator for a type-II down-conversion process for many hours where signal and idler photons can be separated by polarization. Earlier resonator stabilization schemes for comparable single-photon sources were either unsuccessful [171] or the resonator length was scanned [44] limiting the achievable emission rate. The theory for the SRO, developed here from first principles, fits measured count rates and correlation data very well. The single-photon spectrum of the SRO follows the resonator transfer function with a linewidth of only 62 MHz which was measured by evaluating the signal-idler intensity cross-correlation function. This means a linewidth reduction of four to five orders of magnitude compared to the THz width of free-space parametric down-conversion. The excellent agreement between predicted and measured values is summarized by table 4.1.

The generated single photons at 894.3 nm are compatible to the D1 line of cesium. This setup thus establishes a long-term stable source of narrow-band single photons that is suitable for coupling to atomic resonances in future atom-light interfaces. However, its linewidth is still a factor ~ 10 larger than the natural linewidth of the cesium $6^2P_{1/2}$ level. Additionally, many longitudinal modes are emitted which will require the design of a passive filter-cavity [43, 45] that ensures longitudinal single-mode operation. Transversal operation in a TEM_{00} mode can be obtained by fiber-coupled output fields. These issues will be addressed by the double-resonant OPO described in the following section.

Physical Parameter	Predicted Value	Measured Value
SHG Conversion Efficiency	$8.0 \times 10^{-3}/\text{W}$	$3.4 \times 10^{-3}/\text{W}$
SHG Resonator Finesse	145	135
SHG Coupling Efficiency	0.81	0.81
SHG Phase-Matching Temp.	28.3 °C	28.2 °C
SHG Phase-Matching Range	0.9 K	0.75 K
SRO Coupling Efficiency	0.30	0.28
SRO Signal Round-Trip Time	873 ps	867 ps
SRO Pump Finesse	10.0	9.8
SRO Signal Finesse	21.0	18.5
SRO Signal Bandwidth	53 MHz	62 MHz
Signal-Idler Coincidence Rate	694 counts/(s mW)	660 counts/(s mW)
Signal Rate at 1.5 mW pump	5.9×10^4 counts/s	6.7×10^4 counts/s
Idler Rate at 1.5 mW pump	2.9×10^5 counts/s	2.9×10^5 counts/s

Table 4.1: Comparison between predicted and measured parameters for the SRO experiment.

Chapter 5

Narrow-Band Single Photons from a Double-Resonant OPO

The previous chapter introduced basic concepts of nonlinear processes and showed their application to generate narrow-band single photons from an OPO far below threshold. By resonating the signal field, the parametric fluorescence was redistributed into frequencies resonant to the OPO cavity, but the total photon generation rate stayed constant. The deflection of the idler wave out of the resonator required intra-cavity optical elements, thereby reducing the cavity finesse and setting a lower bound to the achievable linewidth. It has been shown [44, 172] that a double-resonant OPO (DRO) can overcome these issues. However, neither active cavity stabilization for reliable high count rates nor proof of the single-photon character by post-selection techniques have been incorporated in any previous realization. This chapter presents the experimental demonstration of a DRO exhibiting these features. Together with the filter setup, introduced in the following chapter 6, they enable the selection of a single longitudinal DRO output mode in order to interface this source with an atomic quantum memory.

5.1 New OPO Design

The DRO aims at the enhancement of both signal and idler fields inside an optical resonator. Since signal and idler photons, generated by frequency-degenerate parametric down-conversion, can be separated by polarization only, a type-II process is required. For negligible walk-off and collinear propagation of signal and idler fields along the resonator symmetry axis, a periodically poled crystal needs to be selected; here, potassium titanyl phosphate (KTP) offers the highest nonlinear coefficient for down-conversion at the se-

71

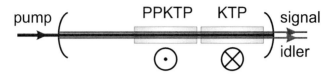

Figure 5.1: Scheme of the double-resonant OPO using a type-II down-conversion process: A compensating unmatched KTP crystal, turned by 90°, accompanies the PPKTP crystal inside the resonator for equal FSR of signal and idler.

lected cesium D1 line (894.3 nm). Since KTP is a biaxial crystal, a second unmatched KTP crystal – turned by 90° – is integrated into the resonator next to the periodically-poled conversion crystal (PPKTP) in order to compensate for the different effective crystal lengths for orthogonally polarized signal and idler fields and to ensure simultaneous resonance [44]. With these boundary conditions in mind, the DRO design will be discussed in the following.

5.1.1 Crystal Parameters

KTP belongs to point group mm2 [144] and exhibits high nonlinear coefficients up to 10.7 pm/V, a high damage threshold, and a wide transparency range $(350 - 4500$ nm). Absorption coefficients are around 9.0 m^{-1} for the pump (447.15 nm) and 0.1 m^{-1} for the fluorescence wavelength. Unfortunately, references for Sellmeier equations are not consistent; data from [173] will be used to determine temperature independent indices of refraction and [174] for temperature dependant expressions.

The PPKTP conversion crystal was ordered with a $\Lambda = 20.8~\mu$m grating period along the x-axis $(\theta = \pi/2, \phi = 0)$. Sellmeier equations yield a spectral width of down-converted fluorescence of ~ 61 GHz (figure 5.2(a)). With an expected FSR$_{S/I}$ for signal and idler modes of ~ 1.5 GHz, this yields an effective number of contributing longitudinal resonator modes < 50. The temperature acceptance range is 2.9 K around the phase-matching temperature of 13.5 °C for frequency-degenerate output (figure 5.2(b)). This acceptance range exceeds the values of most other nonlinear crystals. Still, special care needs to be taken to stabilize the crystal temperature since fluctuations by only $\pm 500~\mu$K result in a frequency jitter of the signal and idler modes of several MHz.

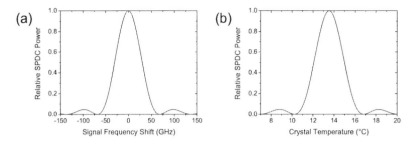

Figure 5.2: (a) Dependance of down-converted power on signal frequency shift to evaluate the expected number of longitudinal resonator modes that may contribute to the nonlinear process. (b) The output power variation with crystal temperature shows a wide acceptance range of 2.9 K, and frequency degeneracy at around 13.5°C can be derived. Still, stability requirements are tough in order to minimize the frequency jitter of signal and idler frequencies.

For type-II phase-matching along the x-axis, an effective nonlinear coefficient

$$d_{eff} = \frac{2}{\pi} d_{24} \sin \theta = 2.32 \; \frac{\text{pm}}{\text{V}}$$

is calculated according to equation 4.2 with $d_{24} = 3.65$ pm/V. Thereby, pump and signal polarizations are assumed parallel to the y-axis, idler polarization parallel to the z-axis.

PPKTP and KTP crystals must be of equal length to compensate for the propagation delay between signal and idler in the conversion crystal. A crystal length of 20 mm each was chosen for high single-pass conversion efficiency. The pump waist inside the crystal will follow design considerations of the DRO resonator since the variation of the Boyd-Kleinman factor for this periodically poled crystal without walk-off is negligible and the smallest possible resonator length will be chosen in order to achieve an *a priori* low number of longitudinal resonator modes.

5.1.2 DRO Resonator

The resonator is designed for simultaneous resonance of pump, signal, and idler fields. Since the PPKTP conversion crystal is preferably placed in the resonator center, a total mirror distance of 68 mm is the smallest experimentally accessible length. With mirror curvatures of 50 mm, an optimal beam waist of 60.7 μm is derived from the resonator ray transfer matrix. This waist results in a Boyd-Kleinman factor of about 0.91 and a single-pass conversion efficiency $\kappa_{NL} = 3.9 \times 10^{-3}$ W^{-1}. Incorporating intra-cavity pump losses, an input mirror reflectivity of ~ 0.48 ensures impedance matching with an expected finesse $F_P^{th} = 4.2$. Thus, a pump power of 1 mW outside the resonator leads to 1.9 mW circulating power. This value itself may not justify the implementation of pump enhancement, but the resonant pump will allow length stabilization of the resonator, as already demonstrated in the SRO setup.

The input mirror is highly reflective at signal and idler wavelengths, and a finesse value of $F_{S/I}^{th} = 272$ is calculated for signal and idler with an output mirror reflectivity of 0.99. This should yield a single-photon bandwidth $\gamma_S = \gamma_I \approx 5.4$ MHz, an improvement of more than an order of magnitude compared to the SRO realization. The effective cavity lengths for pump and fluorescent fields correspond to a free spectral range FSR$_P \approx 1.45$ GHz and FSR$_{S/I} \approx 1.50$ GHz, respectively.

5.2 Estimates for Threshold and Count Rate

The DRO threshold can be obtained classically as introduced in section 4.2. Single-pass conversion gain and losses cancel at threshold; thus, the DRO threshold power P_{thres} may be calculated by the conversion efficiency κ_{NL} and signal round-trip losses. An expected loss of $\epsilon_S = \epsilon_I = 0.011$ leads to

$$P_{thres} \approx 91.4 \text{ mW}.$$

In order to derive signal and idler count rates, a quantum mechanical approach is necessary. Since the full theoretical treatment of the multi-mode DRO, which will be subject of subsection 5.3, is quite complex, a quantum mechanical description of a single-mode double-resonant OPO via a Fokker-Planck ansatz [175] will shortly be sketched in the following. It allows the calculation of the count rates for single-mode output only, though, and can thus just estimate the order of magnitude. The multi-mode output can then be estimated by scaling with the number of effectively contributing resonator modes.

In order to obtain c-number representations α_k $(k = S, I)$ for the intra-cavity mode operators a_k at pump power P, the corresponding stochastic differential equations are linearized and solved by Fourier transform. Equation 5.7 of [175] provides the spectral information

$$\langle \alpha_j^\dagger(\omega')\alpha_k(\omega)\rangle = \left(\frac{s_{jk}(\omega)}{2\sqrt{\gamma_j\gamma_k}}\right)\delta(\omega + \omega')$$

where

$$s_{SS}(\omega) = s_{II}(\omega) = \varepsilon \left(\frac{1}{(1-\varepsilon)^2 + \bar{\omega}^2} - \frac{1}{(1+\varepsilon)^2 + \bar{\omega}^2} \right)$$

with $\varepsilon = \sqrt{P/P_{thres}}$ and $\bar{\omega} = \omega/\gamma_{S/I}$. The count rates then follow as

$$
\begin{aligned}
R_k &= \eta \langle a_{k,out}^\dagger(t) a_{k,out}(t) \rangle = 2\pi \gamma_k \eta \langle a_k^\dagger(t) a_k(t) \rangle \\
&= \frac{2\pi \gamma_k \eta}{2\pi} \int d\omega \langle \alpha_k^\dagger(\omega') \alpha_k(\omega) \rangle \\
&= \frac{2\pi \gamma_k \eta \, \varepsilon^2}{2 \left(1 - \varepsilon^2\right)}.
\end{aligned}
$$

Here, $a_{k,out}$ are the mode operators outside the cavity and $\eta \approx 0.44$ the cavity escape efficiency. An effective number $N = 45$ of resonator modes within the phase-matching sinc-function and a setup transmission of 0.13 for the idler due to APD detection efficiency and spectral filters is calculated. With these parameters and 1 mW pump power, a multi-mode count rate of $\approx 9.6 \times 10^5$ Hz at the idler APD is expected.

5.3 Theoretical Description of the DRO

The exact expressions for count rates, spectral information, and signal-idler cross-correlation function of the DRO are developed analogously to the corresponding SRO equations[1]. Thus, their derivation will be sketched only briefly. First, the signal and idler operators for free space and inside the crystal are calculated. Then the bi-photon wave function can be deduced from the interaction Hamiltonian. Finally, experimentally accessible parameters like count rates and correlation functions follow. All parameters are defined as introduced in section 4.5.

Field operators: The free-field operator of the signal mode reads

$$E_S^{(+)}(x,t) = \sqrt{\frac{\hbar \omega_S}{2\epsilon_0 c A}} \int_{-\infty}^{\infty} \frac{d\Omega}{\sqrt{2\pi}} a_S(\omega_S + \Omega) e^{i(\omega_S + \Omega)\left(\frac{x}{c} - t\right)}$$

with

$$\left[a_S(\omega), a_S^\dagger(\omega') \right] = \delta(\omega - \omega').$$

A corresponding expression is valid for the idler mode:

$$E_I^{(+)}(x,t) = \sqrt{\frac{\hbar \omega_I}{2\epsilon_0 c A}} \int_{-\infty}^{\infty} \frac{d\Omega}{\sqrt{2\pi}} a_I(\omega_I + \Omega) e^{i(\omega_I + \Omega)\left(\frac{x}{c} - t\right)}$$

---

[1]All frequencies in this particular section are denoted as angular frequencies.

with

$$\left[a_I(\omega), a_I^\dagger(\omega') \right] = \delta(\omega - \omega').$$

Since signal and idler photons are orthogonally polarized, $[a_S(\omega), a_I^\dagger(\omega')] = 0$ holds.

In order to find the field operators inside the crystal, the longitudinal signal mode frequencies

$$\omega_{S,m_S} = \frac{(m_{S,0} + m_S)\pi c}{n_s(\omega_{S,m_S})l}$$

are used with

$$m_{S,0} = \frac{\omega_S n_S l}{\pi c},$$

$n_S = n_s(\omega_S)$, $m_S \in \mathbb{Z}$, and $m_{S,0} \gg |m_S|$.

The cavity can be assumed on resonance since the total spectral envelope of the signal due to phase-matching is much larger than the mode spacing, i.e. $m_{S,0} \in \mathbb{N}$. Using Taylor expansion

$$n_s(\omega_{S,m_S}) = n_S + (\omega_{S,m_S} - \omega_S) \frac{\partial n_s(\omega)}{\partial \omega}|_{\omega = \omega_S},$$

one finds

$$\omega_{S,m_S} = \omega_S + m_S \frac{\pi c}{\left(n_S + \omega_S \frac{\partial n_s(\omega)}{\partial \omega}|_{\omega = \omega_S} \right) l} = \omega_S + m_S \Delta\omega_{S,cr}.$$

Thus, the signal field operator inside the crystal can be written as

$$E_S^{cr,(+)}(x,t) = \sqrt{\frac{\hbar \omega_S \Delta\omega_{S,cr}}{\epsilon_0 n_S c A \pi}} \sum_{m_S=-\infty}^{\infty} a_{S,m_S} \frac{e^{i\omega_{S,m_S}\left[\frac{x}{c} n_s(\omega_{S,m_S}) - t \right]}}{2}$$

Accounting for resonator losses, the a_{S,m_S} turn into time-dependant quasi-modes

$$a_{S,m_S}(t) = \frac{1}{\sqrt{2\pi}} \int_{-\infty}^{\infty} d\Omega \, a_S(\omega_{S,m_S} + \Omega) \frac{\sqrt{\gamma_S}}{\frac{\gamma_S}{2} + i\Omega} e^{-i\Omega t}$$

with the commutation relation of the Fourier components

$$\left[a_S(\omega_{S,m_S} + \Omega), a_S^\dagger(\omega_{S,m_S'} + \Omega') \right] = \delta_{m_S,m_S'} \delta(\Omega - \Omega').$$

The operator for the relevant signal field component in the lossy cavity can then be written as

$$
\begin{aligned}
E_S^{cr,(+)}(x,t) = & \sqrt{\frac{\hbar \omega_S}{2\epsilon_0 n_S c A}} \frac{\sqrt{\gamma_S \Delta\omega_{S,cr}}}{2\pi} \times \\
& \times \sum_{m_S=-\infty}^{\infty} \int_{-\infty}^{\infty} d\Omega \frac{a_S(\omega_{S,m_S} + \Omega)}{\frac{\gamma_S}{2} + i\Omega} e^{i\left[k_{S,m_S}(\Omega)x - (\omega_{S,m_S} + \Omega)t \right]}
\end{aligned}
$$

using

$$k_{S,m_S}(\Omega) = \frac{\omega_{S,m_S} + \Omega}{c} n_s(\omega_{S,m_S} + \Omega)\,.$$

For a resonator length $L_r > l$, one lets $\Delta\omega_{S,cr} \rightarrow \Delta\omega_{S,eff}$ and $\omega_{S,m_S} \rightarrow \omega_S + m_S\Delta\omega_{S,eff}$. Here, $\Delta\omega_{S,eff}$ is the effective free spectral range

$$\Delta\omega_{S,eff} = \frac{2\pi}{T_S} \quad \text{and} \quad T_S = \frac{2l}{v_{g,S}} + \frac{2\bar{n}_S(L_r - l)}{c}$$

with the mean refractive index $\bar{n}_S$ for the resonator space outside the PPKTP crystal.
Equivalent expressions hold for the idler field:

$$\begin{aligned}
E_I^{cr,(+)}(x,t) &= \sqrt{\frac{\hbar\omega_I}{2\epsilon_0 n_I cA}} \frac{\sqrt{\gamma_I\Delta\omega_{I,eff}}}{2\pi} \times \\
&\times \sum_{m_I=-\infty}^{\infty} \int_{-\infty}^{\infty} d\Omega\, \frac{a_I(\omega_{I,m_I} + \Omega)}{\frac{\gamma_I}{2} + i\Omega} e^{i[k_{I,m_I}(\Omega)x - (\omega_{I,m_I}+\Omega)t]} \\
k_{I,m_I}(\Omega) &= \frac{\omega_{I,m_I} + \Omega}{c} n_i(\omega_{I,m_I} + \Omega) \\
\Delta\omega_{I,eff} &= \frac{2\pi}{T_I} \quad \text{with} \quad T_I = \frac{2l}{v_{g,I}} + \frac{2\bar{n}_I(L_r - l)}{c}\,.
\end{aligned}$$

Bi-photon wave function: The macroscopic pump field can be described by a classical wave

$$E_P^{cr}(x,t) = E_P e^{i[k_P(\omega_P)x - \omega_P t]}\,,$$

and the phase-matching conditions for the nonlinear interaction between pump, signal, and idler read

$$\begin{aligned}
\omega_P &= \omega_S + \omega_I\,, \\
\vec{k}_P(\omega_P) &= \vec{k}_S(\omega_S) + \vec{k}_I(\omega_I)\,.
\end{aligned}$$

The interaction Hamiltonian describing down-conversion has the form

$$H_{int} = \frac{\chi}{2l}\int_{-l}^{0} dx \left(E_P^{cr} E_S^{cr,(-)} E_I^{cr,(-)} + E_P^{cr*} E_S^{cr,(+)} E_I^{cr,(+)} \right)\,.$$

Using $\omega_P = \omega_S + \omega_I$, this yields after some minor algebra

$$\begin{aligned}
H_{int} &= i\hbar\alpha \sum_{m_S=-\infty}^{\infty}\sum_{m_I=-\infty}^{\infty} \int_{-\infty}^{\infty} d\Omega \int_{-\infty}^{\infty} d\Omega'\, \frac{\sqrt{\gamma_S\gamma_I}}{\left(\frac{\gamma_S}{2} - i\Omega\right)\left(\frac{\gamma_I}{2} - i\Omega'\right)} \times \\
&\times F_{m_S,m_I}(\Omega,\Omega')\, a_S^\dagger(\omega_{S,m_S} + \Omega)\, a_I^\dagger(\omega_{I,m_I} + \Omega') \times \\
&\times \exp\left(i(m_S\Delta\omega_{S,eff} + \Omega + m_I\Delta\omega_{I,eff} + \Omega')t\right) + \text{h.a.}
\end{aligned}$$

where

$$\alpha = \frac{-iE_P}{16\pi^2\epsilon_0 cA}\sqrt{\frac{\omega_S\omega_I}{n_S n_I}}\chi(\omega_P;\omega_S,\omega_I)\sqrt{\Delta\omega_{S,eff}\Delta\omega_{I,eff}}$$

and

$$
\begin{aligned}
F_{m_S,m_I}(\Omega,\Omega') &= \frac{\chi(\omega_P;\omega_{S,m_S}+\Omega,\omega_{I,m_I}+\Omega')}{\chi(\omega_P;\omega_S,\omega_I)l} \times \\
&\quad \times \int_{-l}^{0} dx\, e^{i\left(k_P-k_{S,m_S}(\Omega)-k_{I,m_I}(\Omega')\right)x} \\
&\overset{\chi\approx\text{const.}}{=} \operatorname{sinc}\left[\frac{l}{2c}\left\{(m_S\Delta\omega_S+\Omega)\left(n_S+\omega_S\frac{\partial n_s}{\partial\omega}|_{\omega=\omega_S}\right) + \right.\right. \\
&\quad \left.\left. + (m_I\Delta\omega_I+\Omega')\left(n_I+\omega_I\frac{\partial n_i}{\partial\omega}|_{\omega=\omega_I}\right)\right\}\right] \times \\
&\quad \times \exp\left(\frac{il}{2c}\left\{(m_S\Delta\omega_S+\Omega)\left(n_S+\omega_S\frac{\partial n_s}{\partial\omega}|_{\omega=\omega_S}\right) + \right.\right. \\
&\quad \left.\left. + (m_I\Delta\omega_I+\Omega')\left(n_I+\omega_I\frac{\partial n_i}{\partial\omega}|_{\omega=\omega_I}\right)\right\}\right).
\end{aligned}
$$

In the regime far below threshold, the bi-photon production rate is much smaller than the signal and idler resonator damping rates ($\kappa \ll \gamma_S,\gamma_I$). Thus, the resonator can be assumed empty before each conversion event. The nonlinear interaction is then treated perturbatively with an initial signal-idler field $|0\rangle = |0\rangle_S \otimes |0\rangle_I$ at time $t = 0$.
For times $t = \Delta t \ll \kappa^{-1}$, the combined signal-idler field becomes $|\psi(\Delta t)\rangle \propto |0\rangle + |\tilde{\psi}(\Delta t)\rangle$ with

$$
\begin{aligned}
\left|\tilde{\psi}(\Delta t)\right\rangle &= \frac{1}{i\hbar}\int_0^{\Delta t} dt\, H_{int}(t)|0\rangle \\
&= \sum_{m_S=-\infty}^{\infty}\sum_{m_I=-\infty}^{\infty}\int_{-\infty}^{\infty} d\Omega \int_{-\infty}^{\infty} d\Omega'\, \frac{\alpha\sqrt{\gamma_S\gamma_I}}{\left(\frac{\gamma_S}{2}-i\Omega\right)\left(\frac{\gamma_I}{2}-i\Omega'\right)} \times \\
&\quad \times \Delta t\operatorname{sinc}\left[\frac{1}{2}\left(m_S\Delta\omega_{S,eff}+\Omega+m_I\Delta\omega_{I,eff}+\Omega'\right)\Delta t\right] \times \\
&\quad \times F_{m_S,m_I}(\Omega,\Omega')\, e^{\frac{i}{2}(m_S\Delta\omega_{S,eff}+\Omega+m_I\Delta\omega_{I,eff}+\Omega')\Delta t} \times \\
&\quad \times a_S^\dagger(\omega_{S,m_S}+\Omega)\, a_I^\dagger(\omega_{I,m_I}+\Omega')|0\rangle
\end{aligned}
$$

Due to the properties of the sinc-function, the integral is dominated by the region $m_S\Delta\omega_{S,eff} + \Omega + m_I\Delta\omega_{I,eff} + \Omega' \leq \pi/\Delta t$. At sufficiently large Δt, the following substitution is valid:

$$\Delta t\operatorname{sinc}\left[\frac{1}{2}(\dots)\Delta t\right] e^{\frac{i}{2}(\dots)\Delta t} \to \pi\,\delta(\dots).$$

This leads to the final expression for the bi-photon wave function

$$|\psi\rangle = \sum_{m_S=-\infty}^{\infty} \sum_{m_I=-\infty}^{\infty} \int_{-\infty}^{\infty} d\Omega \int_{-\infty}^{\infty} d\Omega' \frac{\pi\alpha\sqrt{\gamma_S\gamma_I}}{\left(\frac{\gamma_S}{2} - i\Omega\right)\left(\frac{\gamma_I}{2} - i\Omega'\right)} \times$$
$$\times F_{m_S,m_I}(\Omega, \Omega') \times \delta\left(m_S\Delta\omega_{S,eff} + \Omega + m_I\Delta\omega_{I,eff} + \Omega'\right) \times$$
$$\times a_S^\dagger(\omega_{S,m_S} + \Omega)\, a_I^\dagger(\omega_{I,m_I} + \Omega')|0\rangle\,.$$

Bi-photon generation rate: Once the bi-photon wave function is known, the photon generation rate

$$\kappa = \eta\,\langle\psi|\psi\rangle$$

can be calculated where η represents the cavity escape efficiency. Using operator commutation relations, one obtains

$$\kappa = \eta \sum_{m_S,m_I,m_S',m_I'=-\infty}^{\infty} \int_{-\infty}^{\infty} d\Omega \times$$

$$\times \frac{\pi^2|\alpha|^2\gamma_S\gamma_I}{\left(\frac{\gamma_S}{2} + i\left(\Omega - m_S'\Delta\omega_{S,eff}\right)\right)\left(\frac{\gamma_S}{2} - i\left(\Omega - m_S\Delta\omega_{S,eff}\right)\right)} \times$$

$$\times \frac{\mathrm{sinc}^2\left[\frac{l\Omega}{2c}\left(n_S + \omega_S\frac{\partial n_s}{\partial\omega}|_{\omega=\omega_S} - n_I - \omega_I\frac{\partial n_i}{\partial\omega}|_{\omega=\omega_I}\right)\right]}{\left(\frac{\gamma_I}{2} - i\left(\Omega + m_I'\Delta\omega_{I,eff}\right)\right)\left(\frac{\gamma_I}{2} + i\left(\Omega + m_I\Delta\omega_{I,eff}\right)\right)}$$

Defining $\Delta n := n_S + \omega_S\frac{\partial n_s}{\partial\omega}|_{\omega=\omega_S} - n_I - \omega_I\frac{\partial n_i}{\partial\omega}|_{\omega=\omega_I}$ and $\tau_0 = \frac{l\Delta n}{c}$, this yields

$$\kappa = \eta\,\pi^2\,|\alpha|^2 \sum_{m_S,m_I,m_S',m_I'=-\infty}^{\infty} \frac{2\pi i}{i\frac{\gamma_S-\gamma_I}{2} + m_S'\Delta\omega_{S,eff} + m_I\Delta\omega_{I,eff}} \times$$

$$\times \left\{ \frac{\gamma_S\,\mathrm{sinc}^2\left[\left(m_S'\Delta\omega_{S,eff} + i\frac{\gamma_S}{2}\right)\frac{\tau_0}{2}\right]}{\left[i\gamma_S + (m_S' - m_S)\Delta\omega_{S,eff}\right]\left[i\frac{\gamma_S+\gamma_I}{2} + m_S'\Delta\omega_{S,eff} + m_I'\Delta\omega_{I,eff}\right]} \right.$$

$$\left. - \frac{\gamma_I\,\mathrm{sinc}^2\left[\left(-m_I\Delta\omega_{I,eff} + i\frac{\gamma_I}{2}\right)\frac{\tau_0}{2}\right]}{\left[i\gamma_I + (m_I' - m_I)\Delta\omega_{I,eff}\right]\left[i\frac{\gamma_S+\gamma_I}{2} - m_S\Delta\omega_{S,eff} - m_I\Delta\omega_{I,eff}\right]} \right\}\,.$$

In the case $\gamma_S = \gamma_I$ and $\Delta\omega_{S,eff} = \Delta\omega_{I,eff}$, the summands with $m_I = m_S'$ are evaluated by taking the limit $m_I \to m_S'$ to avoid singularities.

Output spectra for signal and idler fields: The output spectra are defined as

$$S_{S/I}(\omega) = \frac{1}{2\pi}\int_{-\infty}^{\infty} d\tau\, G_{S/I}^{(1)}(\tau)e^{i\omega\tau}$$

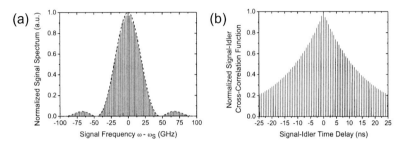

Figure 5.3: (a) Normalized signal spectrum with $\Delta\omega_{S/I} = 1.5$ GHz and $\gamma_{S/I} = 10$ MHz. The propagation delay was set to the PPKTP value of $\tau_0 = 7.3$ ps, yielding a phase-matching envelope of 61 GHz width (dashed line). (b) Normalized signal-idler cross-correlation function using the same parameter set.

where

$$
\begin{aligned}
G_{S/I}^{(1)}(\tau) &= \eta \, \langle\psi|E_{S/I}^{(-)}(x,t)E_{S/I}^{(+)}(x,t+\tau)|\psi\rangle \\
&= \eta \sum_{m_S,m_I,m'_S,m'_I=-\infty}^{\infty} \int_{-\infty}^{\infty} d\Omega \, \frac{\hbar\omega_S\pi|\alpha|^2}{4\epsilon_0 cA} \times \\
&\times \frac{\gamma_S\gamma_I}{\left(\frac{\gamma_S}{2}+i(\Omega-m'_S\Delta\omega_{S,eff})\right)\left(\frac{\gamma_S}{2}-i(\Omega-m_S\Delta\omega_{S,eff})\right)} \times \\
&\times \frac{\mathrm{sinc}^2\left(\Omega\frac{\tau_0}{2}\right)e^{-i(\omega_S+\Omega)\tau}}{\left(\frac{\gamma_I}{2}-i(\Omega+m'_I\Delta\omega_{I,eff})\right)\left(\frac{\gamma_I}{2}+i(\Omega+m_I\Delta\omega_{I,eff})\right)}
\end{aligned}
$$

Comparing this result with the spectrum definition leads to

$$
S_S(\omega) = 2\pi\,\eta\,\frac{\hbar\omega_S|\alpha|^2}{8\epsilon_0 cA}\left|\sum_{m_S=-\infty}^{\infty}\sum_{m_I=-\infty}^{\infty}\times\right. \tag{5.1}
$$

$$
\times\left.\frac{\sqrt{\gamma_S\gamma_I}\,\mathrm{sinc}\left[\frac{\tau_0}{2}(\omega-\omega_S)\right]}{\left(\frac{\gamma_S}{2}-i\left((\omega-\omega_S)-m_S\Delta\omega_{S,eff}\right)\right)\left(\frac{\gamma_I}{2}+i\left((\omega-\omega_S)+m_I\Delta\omega_{I,eff}\right)\right)}\right|^2
$$

for the signal field. A normalized signal spectrum is depicted in figure 5.3(a); the spectral width of phase-matching sinc-envelope is set to 61 GHz.

The following form of the signal spectrum is favorable in order to calculate the FWHM of individual lines:

$$
S_S(\omega) = 2\pi\,\eta\sum_{m_S,m_I,m'_S,m'_I=-\infty}^{\infty}\frac{\hbar\omega_S|\alpha|^2\gamma_S\gamma_I}{8\epsilon_0 cA}\mathrm{sinc}^2\left[\frac{\tau_0}{2}(\omega-\omega_S)\right]\times
$$

$$\times \frac{\left(\frac{\gamma_S}{2}\right)^2 + (\omega - (\omega_S + m'_S \Delta\omega_{S,eff}))\,(\omega - (\omega_S + m_S \Delta\omega_{S,eff}))}{\left(\left(\frac{\gamma_S}{2}\right)^2 + (\omega - (\omega_S + m'_S \Delta\omega_{S,eff}))^2\right)\left(\left(\frac{\gamma_S}{2}\right)^2 + (\omega - (\omega_S + m_S \Delta\omega_{S,eff}))^2\right)} \times$$

$$\times \frac{\left(\frac{\gamma_I}{2}\right)^2 + (\omega - (\omega_S - m'_I \Delta\omega_{I,eff}))\,(\omega - (\omega_S - m_I \Delta\omega_{I,eff}))}{\left(\left(\frac{\gamma_I}{2}\right)^2 + (\omega - (\omega_S - m'_I \Delta\omega_{I,eff}))^2\right)\left(\left(\frac{\gamma_I}{2}\right)^2 + (\omega - (\omega_S - m_I \Delta\omega_{I,eff}))^2\right)} \, .$$

The FWHM is then derived as

$$\mathrm{FWHM} \;=\; \sqrt{\frac{\sqrt{\gamma_I^4 + 6\gamma_S^2\gamma_I^2 + \gamma_S^4} - (\gamma_S^2 + \gamma_I^2)}{2}} \; \overset{\gamma_S = \gamma_I}{=} \; \gamma_S\sqrt{\sqrt{2}-1} \approx 0.64\,\gamma_S \, .$$

Signal-idler intensity cross-correlation function: Analogously to the spectrum, the second-order cross-correlation function between signal and idler fields is defined as

$$G_{S/I}^{(2)}(\tau) \;=\; \langle\psi| E_I^{(-)}(x,t) E_S^{(-)}(x,t+\tau) E_S^{(+)}(x,t+\tau) E_I^{(+)}(x,t) |\psi\rangle$$

$$= \left| \sum_{m_S, m_I} \frac{\hbar\alpha\sqrt{\gamma_S\gamma_I\omega_S\omega_I}}{4\epsilon_0 cA} \times \right.$$

$$\left. \times \begin{cases} \dfrac{e^{-\left(\frac{\gamma_S}{2}+im_S\Delta\omega_{S,eff}\right)\left(\tau-\frac{\tau_0}{2}\right)} \mathrm{sinc}\left[\frac{i\tau_0}{2}\left(\frac{\gamma_S}{2}+im_S\Delta\omega_{S,eff}\right)\right]}{\frac{\gamma_S+\gamma_I}{2}+i\left(m_S\Delta\omega_{S,eff}+m_I\Delta\omega_{I,eff}\right)} & \forall\,\tau \geq \frac{\tau_0}{2} \\[2em] \dfrac{e^{\left(\frac{\gamma_I}{2}+im_I\Delta\omega_{I,eff}\right)\left(\tau-\frac{\tau_0}{2}\right)} \mathrm{sinc}\left[\frac{i\tau_0}{2}\left(\frac{\gamma_I}{2}+im_I\Delta\omega_{I,eff}\right)\right]}{\frac{\gamma_S+\gamma_I}{2}+i\left(m_S\Delta\omega_{S,eff}+m_I\Delta\omega_{I,eff}\right)} & \forall\,\tau < \frac{\tau_0}{2} \end{cases} \right|^2 .$$

The normalized $G_{S/I}^{(2)}$ function is shown in figure 5.3(b) for infinite time resolution of the measurement apparatus. Realistic measurements, however, need to take into account the limited time resolution Δt of the HBT setup which can be modeled by a convolution of the $G_{S/I}^{(2)}$ function with a Gaussian

$$w(t) = \sqrt{\frac{4\log 2}{\pi\Delta t^2}} \exp\left[-4\log 2\left(\frac{t}{\Delta t}\right)^2\right] .$$

The modified cross-correlation then reads

$$G_{S/I}^{(2),\mathrm{mod}}(\tau) = \int_{-\infty}^{\infty} dt\, w(\tau - t)\, G_{S/I}^{(2)}(t) . \tag{5.2}$$

Convoluted functions $G_{S/I}^{(2),\mathrm{mod}}$ are provided in figure 5.4 for time resolutions $\Delta t = 400$ ps and $\Delta t = 700$ ps, respectively.

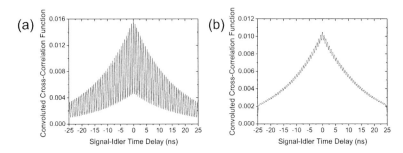

Figure 5.4: Normalized signal-idler cross-correlation functions convoluted with Gaussians that account for the limited time resolution (a) $\Delta t = 400$ ps and (b) $\Delta t = 700$ ps of the HBT setup. The functions were calculated with realistic parameters $\tau_0 = 7.3$ ps, $\Delta\omega_S = \Delta\omega_I = 1.5$ GHz, and $\gamma_S = \gamma_I = 10$ MHz.

5.4 DRO Setup and Cavity Characterization

Section 5.1 introduced the basic concepts to realize double resonance for signal and idler fields in an OPO below threshold. An equal FSR for the two generated fluorescent fields is achieved by an additional KTP crystal that compensates for the propagation delay between orthogonally polarized signal and idler fields in the PPKTP conversion crystal. The DRO pump wave at twice the frequency of the cesium D1 line is generated by SHG analogously to the SRO setup described in section 4.4.

Figure 5.5 depicts a photograph incorporating the principal design issues of the DRO. The pump beam enters an air-tight case through an AR coated window which provides substantial protection against acoustic noise and temperature fluctuations. The cavity mirror mounts are attached to a ground plate that is temperature-stabilized within ± 25 mK by four peltier elements (not shown). This ground plate also holds the tilt aligners which carry one crystal each. The crystals themselves are temperature stabilized within ± 500 μK inside a copper heat sink by peltier elements. The 20 mm long PPKTP crystal is centered inside the cavity, with a spacing of only 2 mm to the KTP crystal. Since tilt aligners with sufficient precision were available with a minimum length of 60 mm only, one of the cavity mirrors must be mounted to an aluminum cylinder; the second mirror is attached to a PZT in order to stabilize the cavity length. Signal and idler fields leave the setup via a second window, AR coated for 894 nm.

As in the SRO realization, length stabilization is provided by the Hänsch-

Outer Case   Tilt Aligner   Ground Plate

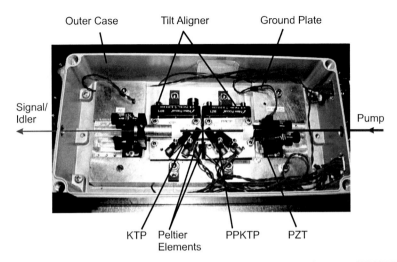

Signal/
Idler   Pump

KTP   Peltier   PPKTP   PZT
       Elements

Figure 5.5: Photograph of the DRO setup inside an air-tight case. PPKTP conversion and KTP compensation crystals are mounted on top of separate tilt aligners with their temperature stabilized individually. The resonator length is stabilized to the pump beam via a PZT.

Couillaud method. A setup consisting of PBS, QWP, and two photo diodes detect the polarization ellipticity of the pump beam reflected from the input mirror. The input mirror reflectivity is chosen to 0.48 for impedance-matching due to the high losses in KTP at 450 nm. The output mirror is highly reflective at this wavelength. Intra-cavity pump parameters can be evaluated by means of the relative power coupled into the resonator when the cavity length is modulated by the PZT. Figure 5.6(a) shows the reflected pump signal at an incident power of 5 mW. Assuming an $\mathrm{FSR}_P \approx 1.45$ GHz for the pump field due to an effective cavity length of 0.103 m, it reveals a linewidth of $\delta\nu_p \approx 375$ MHz, resulting in a finesse of $F_P^{exp} \approx 3.87$. This coincides well with the theoretical estimation in subsection 5.1.2. Impedance- and mode-matching assure a high coupling efficiency of > 0.85. In figure 5.6(b), error signals for cavity stabilization are depicted. They show a high signal-to-noise ratio of $> 100 : 1$, and the cavity can be locked for hours until drifts exceed the PZT travel range.

The signal emitted by the DRO is coupled into a single-mode fiber in order to achieve easy mode-matching with subsequent experimental setups. Mode-matching into the fiber is obtained by coupling a counter-propagating beam into the DRO resonator. This method has the advantage that resonator pa-

83

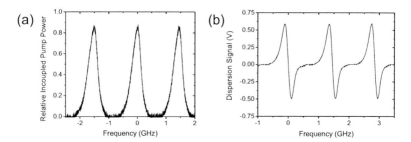

Figure 5.6: (a) Coupling efficiency of the DRO pump beam. A finesse $F_P^{exp} \approx 3.87$ can be derived. (b) Error signal used for cavity stabilization with a signal-to-noise ratio of $100 : 1$.

rameters at the fluorescence wavelength of 894.3 nm can be measured directly (figure 5.7). On the contrary, the signal finesse in the SRO setup had only been determined in an indirect way via signal-idler cross-correlation measurements. From the effective resonator length for signal and idler fields, a free spectral range $FSR_{S/I} \approx 1.50$ GHz can be derived to scale the frequency axis. A resonator linewidth of ≈ 5.7 MHz is determined fitting a Lorentzian which yields a signal finesse $F_S^{exp} \approx 265$. This is close to the theoretical value of $F_S^{th} = 272$. A small residual side maximum remained even after careful adjustment. Its contribution to the total transmission amounts to < 0.1. Its

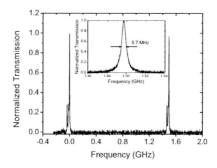

Figure 5.7: Transmission spectrum of the DRO resonator using a counter-propagating beam at the master laser frequency and zoom into the first transmission maximum (inset). A cavity linewidth of 5.7 MHz can be derived.

frequency was changing in time; thus, it is not visible in the inset of figure 5.7. Signal and idler emission pass three long pass filters ($T_{LP} = 0.96$ each) in order to block residual pump light transmitted through the cavity output mirror. They are then split into separate spatial modes by a PBS. The signal APD is fiber-coupled with an efficiency of about 0.15 while the idler is detected free-beam to an APD behind an additional 10 nm wide band pass filter ($T_{BP} = 0.5$). The APD detection efficiency around 900 nm is $\eta_{APD} = 0.35$ as measured in previous experiments. APD dark counts could greatly be reduced to ~ 120 counts/s during laser operation by a thoroughly designed cover.

5.5   Measuring the DRO Output

5.5.1   Frequency Degeneracy and Signal/Idler Count Rates

As depicted in figure 4.12(a), the phase-matching condition of SPDC still allows signal and idler central wavelengths to differ considerably from their frequency-degenerate state which inhibits signal emission on the proper cesium transition. Thus, the PPKTP crystal temperature has to be set carefully in order to achieve both energy and momentum conservation for degenerate output. The KTP temperature is then readjusted for simultaneous resonance of all fields. Width and central frequency of signal and idler spectra are measured by setting the HWP in front of the PBS, separating signal and idler fields, to $\pi/4$. Then, their corresponding projections are coupled into a single-mode fiber and focused onto the slit of a 500 mm spectrograph[2] in a nearby lab. By independently measuring a narrow laser line, its resolution could be determined to 48 pm or 18 GHz.

With changing PPKTP crystal temperature, signal and idler central frequencies shift as depicted in figure 5.8. At 13.5 °C, frequency degeneracy is obtained, and signal as well as idler frequency coincide with the master laser frequency (dashed line in figure 5.8). Obeying energy conservation, the central frequencies move, and a total shift of ~ 806 GHz for the signal field is observed if the temperature reaches 33.1 °C. This is in excellent agreement with the expected value of 809 GHz according to Sellmeier equations. The spectra keep a constant width of about 61 GHz.

Figure 5.9 provides a measurement of the DRO output count rates for signal and idler field, respectively. They show a linear dependance on pump power as predicted by theory for the low excitation regime (sections 5.2 and 5.3).

---

[2]Roper Scientific, SpectraPro 2500i

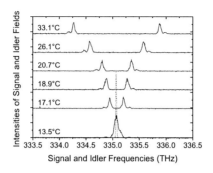

Figure 5.8: Signal and idler central frequencies at different PPKTP crystal temperatures. Due to energy conservation, the output frequencies are emitted symmetrically to the master laser frequency (dashed red line). Frequency degeneracy is achieved for 13.5 °C.

The slopes depend on the setup transmission for the free-beam (idler) and fiber-coupled (signal) port. Taking into account APD detection efficiency, cavity escape efficiency, and filter transmission, the brightness of this source – defined by the count rate behind the PM fiber – can be determined to be 14,000 counts/s per mW pump power and MHz signal bandwidth. This value exceeds all previous realizations of cavity-enhanced parametric down-conversion by a factor of around 200.

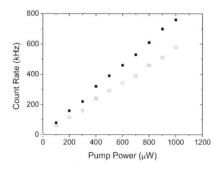

Figure 5.9: Multi-mode count rates for signal ($\square$) and idler ($\blacksquare$) field. The count rates show the expected linear dependance on pump power.

5.5.2 Signal-Idler Cross-Correlations

Measuring signal-idler intensity cross-correlations according to the method described in subsection 4.6.3, one gains considerable information about DRO parameters, and fiber coupling of the APD detecting the signal field assures a lower coherent background. If the DRO is pumped by a TEM_{00}, the spacing of subsequent coincidence peaks, according to a multiple of the DRO round-trip time t_{rt}, can be derived intuitively since signal and idler photon are generated simultaneously. From the exponential decay of the correlation function depicted in figure 5.10(a), cavity decay rates $\gamma_S \approx 4.15$ MHz and $\gamma_I \approx 4.55$ MHz for signal and idler are derived. They translate into single-mode linewidths of 2.7 MHz and 3.0 MHz for signal and idler, respectively. Together with a theoretical $\tau_0 = 7.3$ ps signal-idler propagation delay, these parameters are used to fit the experimental data for the TEM_{00} mode via equation 5.2 with $t_{rt}^{exp} = 686$ ps according to the mirror distance chosen for stable resonator operation. The sums were evaluated with $m_S, m_I \in [-100, \ldots, 100]$ incorporating the effective cavity length, experimental pump power, and the Boyd-Kleinman factor calculated above. An HBT time resolution of 620 ps was assumed, and excellent agreement between theory and experiment could be achieved.

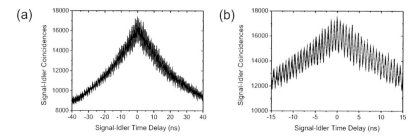

Figure 5.10: (a) Temporal second-order cross-correlation function $G_{S/I}^{(2)}$ between signal and idler modes. Linewidths of 2.7 MHz and 3.0 MHz for signal and idler can be derived from the exponential decays. (b) Zoom into the region around $\tau = 0$ where the red dashed line follows the theoretical model of equation 5.2 that takes into account the time resolution of the correlation setup of ~ 620 ps.

5.5.3   Signal Auto-Correlation

While the signal-idler cross-correlation function reaches values $g_{SI}^{(2)}(0) > 2$, the auto-correlation functions $g_{SS}^{(2)}$ and $g_{II}^{(2)}$ are expected to satisfy $g_{SS}^{(2)}(0) = g_{II}^{(2)}(0) < 2$. For pump powers above the calculated threshold value $P_{thres} \approx$ 91.4 mW, the auto-correlation function at $\tau = 0$ approaches unity, and the DRO emits a coherent state. In the low-excitation regime and for nearly equal signal and idler cavity decay rates, however, it follows [176]

$$ g_{SS/II}^{(2),theo}(0) = 2 \left(\frac{q+2}{q+3} \right)^2 \quad \text{with} \quad q = 2\epsilon_S/\kappa_{NL} - 1 \,. $$

This bunching behavior is a general property of a state generated by many independent uncorrelated emitters. It exhibits an upper limit of $g_{SS/II}^{(2),theo}(0) = 2$ for vanishing conversion efficiency. Figure 5.11 shows a measurement of the temporal auto-correlation function for the signal mode at 1.9 mW pump power. Bunching is observed with $g_{SS}^{(2),exp}(0) = 1.79 \pm 0.4$. Applying the parameters from section 5.2, a theoretical value far below threshold of $g_{SS}^{(2),theo}(0) = 1.76$ is derived which fits the measurement data.

5.5.4   Heralded Single-Photon Statistics

While single-photon generation usually relies on the existence of a two-level system that delays consecutive photon emission by a characteristic reexcitation time, single-photon statistics can also be obtained by projection

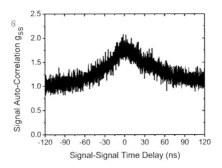

Figure 5.11: Signal auto-correlation function which shows the characteristic bunching phenomenon. The value $g_{SS}^{(2),exp}(0) = 1.79 \pm 0.4$ agrees with the theoretical prediction for a DRO far below threshold.

measurements. The creation of single-photon statistics from SPDC by heralded detection of idler photons has been reported in [177] for the first time. The normalized idler-triggered second-order autocorrelation function, that describes coincidences between an idler and a first signal photon arriving at $t = 0$ as well as a second signal photon detected at $t = \tau$, reads [178]

$$g_c^{(2)}(\tau) = \frac{\left\langle E_S^\dagger(0) E_S^\dagger(\tau) E_S(\tau) E_S(0) \right\rangle_{\text{pm}}}{\left\langle E_S^\dagger(0) E_S(0) \right\rangle_{\text{pm}} \left\langle E_S^\dagger(\tau) E_S(\tau) \right\rangle_{\text{pm}}}$$

where post-selection is taken into account by the idler projection operator $\langle X \rangle_{\text{pm}} = \langle E_I^\dagger(0) X E_I(0) \rangle / \sqrt{\langle E_I^\dagger(0) E_I(0) \rangle}$.
Effectively, small intervals $[-\tau_{Coin}/2, \tau_{Coin}/2]$ and $[\tau - \tau_{Coin}/2, \tau + \tau_{Coin}/2]$ are accepted for the arrival time of a signal photon relative to the heralding idler. Then, the numerator of $g_c^{(2)}$ convolutes a sixth-order moment of field operators with two coincidence windows of width τ_{Coin} at $t = 0$ and $t = \tau$, respectively.
The electrical circuit performing these operations is sketched in figure 5.12. In order to measure single-photon statistics on the signal field, only signal photons within a specified time interval $[-\tau_{Coin}/2, \tau_{Coin}/2]$ around an idler detection event are considered. The edge of an idler APD TTL pulse triggers a monoflop (SN74123, Texas Instruments) with a time interval $\tau_{Coin} \in [25, \ldots, 200]$ ns in the high state. Assuming equal propagation times between DRO resonator and the electronics for signal and idler, the signal APD pulses are delayed by $\tau_{Coin}/2$ to center the coincidence window around an idler detection. The monoflop output window is split, and one of the two coincidence windows is retarded by a variable time τ defining the signal-signal delay. Applying analog multipliers (AD835, Analog Devices), each coincidence window

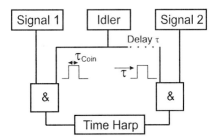

Figure 5.12: Schematic of the electrical circuits for measuring the numerator of the conditional second-order auto-correlation function $g_c^{(2)}$. For details see text.

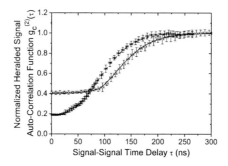

Figure 5.13: Normalized heralded signal auto-correlation function $g_c^{(2)}$ as a function of signal-signal time delay. Antibunching of $g_c^{(2)}(0) \approx 0.2$ and $g_c^{(2)}(0) \approx 0.4$ has been achieved using coincidence windows $\tau_{Coin} = 40$ ns (■) and $\tau_{Coin} = 200$ ns (□), respectively. The dashed line shows a theoretical fit for data taken with $\tau_{Coin} = 40$ ns while the solid line describes a measurement with $\tau_{Coin} = 200$ ns.

becomes mixed with the TTL pulse of a signal APD. The multiplier outputs then feed the start and stop ports of a coincidence counter (TimeHarp 100, PicoQuant).

For each data point $\{\tau, g_c^{(2)}(\tau)\}$ in figure 5.13, signal coincidences arriving at the counter were integrated over $300 - 1000$ s and normalized by the cumulated idler counts within this time interval. Then, the average of several integrations was divided by the corresponding convoluted value of the cross-correlation function $G_{S/I}^{(2)}$, and the result was normalized to obtain $g_c^{(2)}(\infty) = 1$.

The detector time resolution can be neglected compared to the inverse decay rate of the cavity in the MHz regime. The time-dependant part of the denominator is actually the second-order cross-correlation function, again convoluted with the coincidence window. Since the joint state of signal and idler emitted by an OPO far below threshold is a Gaussian state [178], all higher correlation functions can be reduced to second-order moments by the Gaussian moment-factoring theorem. Only $\langle E_k^{\dagger}(t + \tau)E_k(t)\rangle$, $k \in \{S, I\}$, and $\langle E_S^{\dagger}(t + \tau)E_I(t)\rangle$ are non-zero. Explicit expressions have been provided in [179] and are used to fit the experimental data below.

Figure 5.13 shows results for $g_c^{(2)}(\tau)$ at a heralding idler rate $R_H = 100$ kHz. Both $\tau_{Coin} = 40$ ns and $\tau_{Coin} = 200$ ns create antibunching below the characteristic value of 0.5 for a single-photon Fock state. While $g_c^{(2)}(0) \approx 0.2$ is achieved for $\tau_{Coin} = 40$ ns, the convolution with a wider coincidence window

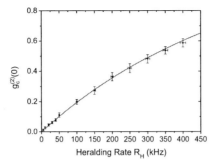

Figure 5.14: Normalized idler-triggered signal auto-correlation function $g_c^{(2)}(0)$. The solid line follows the theoretical model. This parameter characterizes the quality of our heralded single-photon source, and values down to $g_c^{(2)}(0) \approx 0.01$ could be achieved.

$\tau_{Coin} = 200$ ns leads to a less pronounced dip of $g_c^{(2)}(0) \approx 0.4$. Proof of single-photon statistics is still obtained for wide coincidence windows compared to broad-band SPDC due to the long coherence length of signal and idler wave packets produced by the DRO. The theoretical fits follow the theory presented above, assuming a mean cavity decay rate $\gamma_{S/I} = 5.1$ MHz as the only free parameter, and show an excellent agreement with measured data. The dependence of the critical parameter $g_c^{(2)}(0)$ on the heralding rate R_H is depicted in figure 5.14 for a coincidence window width $\tau_{Coin} = 40$ ns, and $g_c^{(2)}(0) = 0.012 \pm 0.005$ has been achieved for $R_H = 5$ kHz. Antibunching below 0.5 can be observed up to $R_H = 300$ kHz.

5.5.5 Discussion

Admixtures of higher photon number $n \geq 2$ in the single-photon state will limit its applicability to tasks in quantum information processing. Thus, the ratio $\beta = P(n \geq 2)/P(1) = P(1) g_c^{(2)}(0)/2$ in the photon number distribution outside the resonator is an important distinguishing parameter of this narrow-band source. $P(1)$ and $P(n \geq 2)$ are determined using the approach introduced in [177], yielding $P(1) = 0.55 \pm 0.01$. The measurements in figure 5.15 show values down to $\beta \approx 3.3 \times 10^{-3}$ at $R_H = 5$ kHz. Compared to a Poissonian source of equal heralding efficiency $P(1)$, higher photon number contributions $P(n \geq 2)$ have been reduced by nearly two orders of magnitude.

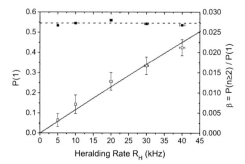

Figure 5.15: Suppression of higher photon number contributions. Measurements of the heralding efficiency $P(1)$ (■) and the relative suppression $\beta = P(n \geq 2)/P(1)$ (□) are plotted versus the heralding rate R_H. $P(1)$ is constant at 0.55 as expected in this regime (dashed line). β follows a linear behavior with the solid line providing a fit according to theory [177].

All theoretical predictions in this chapter are derived form first-principle calculations, and their agreement with experimental measurements is shown in table 5.1.

Physical Parameter	Predicted	Measured
Temperature at Frequency Degeneracy	13.5 °C	13.5 °C
Pump Finesse	4.2	3.9
Pump Decay Rate	343 MHz	375 MHz
Signal Finesse (Cross-Correlations)	272	351
Idler Finesse (Cross-Correlations)	272	320
Signal Decay Rate (Cross-Correlations)	5.4 MHz	4.2 MHz
Idler Decay Rate (Cross-Correlations)	5.4 MHz	4.6 MHz
Signal Finesse (Counter-Propagation)	272	265
Signal Decay Rate (Counter-Propagation)	5.4 MHz	5.7 MHz
Multi-Mode Count Rate at 1 mW pump	960 kHz	760 kHz
Signal/Idler Frequency Shift ($\Delta T = 33.1$ °C)	809 GHz	806 GHz
Signal/Idler Round-Trip Time	667 ps	686 ps
Signal-Signal Bunching $g_{SS}^{(2)}(0)$	1.76	1.79
$g_c^{(2)}(0)$ at $\tau_{Coin} = 40$ ns and $R_H = 100$ kHz	0.19	0.19
$g_c^{(2)}(0)$ at $\tau_{Coin} = 200$ ns and $R_H = 100$ kHz	0.41	0.40
$g_c^{(2)}(0)$ at $\tau_{Coin} = 40$ ns and $R_H = 5$ kHz	0.01	0.01

Table 5.1: Comparison between predicted and measured parameters for the DRO experiment.

Chapter 6

DRO in Single-Mode Operation

*"A designer knows he has achieved perfection
not when there is nothing left to add,
but when there is nothing left to take away."*
– Antoine de Saint-Exupéry

Depending on the application, narrow-band single-photon generation in a
single longitudinal DRO resonator mode may be required. While the proof
of single-photon statistics or selected experiments regarding photon wave
packet non-locality can also be studied in multi-mode operation, further frequency selection is inevitable for coupling to atomic resonances. This chapter
accounts for experimental techniques in order to filter such a single longitudinal mode and proves single-mode operation of the DRO presented before.

6.1 Conceptual Idea

The DRO spectrum consists of a ~ 61 GHz wide phase-matching envelope
with individual modes of ~ 3 MHz width in ~ 1.5 GHz intervals. Thus,
commercial interference band pass filters are not suitable for single-mode
filtering, and an approach via a Fabry-Pérot cavity is required. Accidental
coincidences between DRO resonator modes and Fabry-Pérot side maxima
will need to be suppressed by a second filter stage that consists of an etalon.
The solid line in figure 6.1 depicts the DRO comb spectrum according to
equation 5.1. Selecting a mirror distance of 4.5 cm and a reflectivity of 0.96,
a free spectral range $\text{FSR}_{FP} = 3.3$ GHz and a linewidth $\Delta \nu_{FP}^{theo} = 43$ MHz

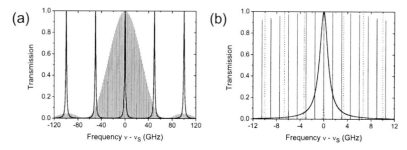

Figure 6.1: (a) Simulated transmission of the DRO output modes through the filter setup. The thin solid line shows the DRO spectrum within the phase-matching envelope while the thick solid line follows the etalon transmission. (b) Zoom into to central DRO mode that is transmitted by the proposed setup. The dotted line depicts the Airy function of the filter cavity.

for this filter cavity can be achieved. The corresponding Airy function is depicted as the dashed line in figure 6.1(b). Finally, the etalon parameters are chosen to 3 mm thickness and 0.9 reflectivity, yielding $FSR_{Et} = 50.0$ GHz and $\Delta\nu_{Et}^{theo} = 1.7$ GHz. As shown in the same figure, the etalon effectively suppresses spurious emission of DRO modes through the filter cavity.

The etalon transmission can be stabilized passively by temperature and a PZT spacer. The filter cavity, on the other hand, consists of two individual mirrors, and active stabilization is required. In a first approach, one may create a control signal for length stabilization by using transmitted DRO photons. However, this will result in a bad signal-to-noise ratio and, moreover, destroy the filtered state. A counter-propagating portion of the master laser may provide a reference signal for filter cavity stabilization, but back reflections are not sufficiently attenuable by polarization optics and will contaminate the DRO state that travels at the same frequency. The solution is a reference lock beam at a wavelength of 852.1 nm that considerably differs from the master laser and DRO signal wavelength at 894.3 nm and can thus be blocked by interference band pass filters. Unfortunately, transmission maxima of signal and reference beams will not coincide in general. Therefore, a mediating stabilization setup, the so-called *transfer setup*, is used to lock the 852.1 nm reference laser[1] to the master laser at an appropriate frequency offset for simultaneous resonance in the filter cavity.

---

[1]Toptica Photonics, DL PRO

6.2 Experimental Realization

6.2.1 Transfer Setup

The stabilization of the master laser is transferred to the reference laser by an intermediary Fabry-Pérot resonator. This setup was built inside an isolated box that limits acoustic perturbations. All optical components use 1/2" optics and have been minimized in size in order to shift mechanical resonances to higher frequencies. The ground plate is mounted on top of four peltier elements that keep its temperature within ±25 mK to avoid drifts due to thermal expansion.

Figure 6.2 shows a scheme of this transfer setup whose cavity parameters coincide with those of the filter cavity introduced in section 6.1. A photograph of the transfer setup is depicted in figure B.3. The master laser is mode-matched to this cavity, and its mirror distance is locked to the master frequency via the Pound-Drever-Hall (PDH) technique [152, 153, 180] using a PZT. Sidebands are created by phase-modulation via an EOM, and the reflected beam from the cavity input mirror is deflected onto a photo diode (PD) by a 50 : 50 BS. The modulation frequency is chosen around $\Omega_M \approx 6$ MHz which means operation of the PDH method in the regime of low modulation frequency $\Omega_M \ll \Delta\nu_{FP}$. The first mechanical PZT resonance is around 2 kHz which limits the lock bandwidth.

In an equivalent PDH setup, the reference laser is stabilized to the transfer cavity. Mostly long-term drifts between master and reference lasers need to be accounted for. Figure 6.3 depicts transmission and error signals of the master laser when the cavity PZT is scanned over one FSR. Analogous measurements have been performed for the reference laser in order to achieve

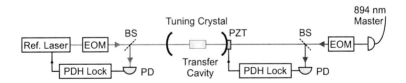

Figure 6.2: Scheme of the transfer setup. The cavity is stabilized to the master laser by the PDH technique. The reference laser, on the other hand, is locked to this cavity and thus follows the frequency deviation of the master. A birefringent crystal inside the cavity allows continuous tuning of the frequency offset between the orthogonally polarized master and reference lasers via crystal temperature. For details see text.

97

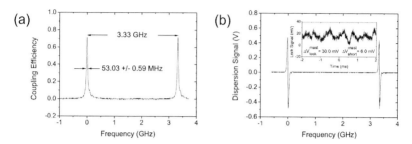

Figure 6.3: With the transfer cavity PZT scanned over one FSR, the cavity transmission spectrum and the error signal of the PDH lock for the master laser are recorded. (a) From the transmission spectrum, a cavity linewidth of ~ 53 MHz is derived. (b) Together with the data for the reference laser, error and lock signals give information about the expected lock accuracy.

high coupling efficiencies and a good signal-to-noise ratio of the error signals. In general, a coupling efficiency > 0.8 and a signal-to-noise ratio of $\approx 200 : 1$ can be achieved which allows a stable lock over hours until drifts exceed the PZT travel range. The inset of figure 6.3(b) shows the lock signal. While the short-term linewidth of the master laser induces a voltage noise of $\Delta V_{short}^{mast} = 6.0$ mV, the lock signal deviates with a peak-peak amplitude of $\Delta V_{lock}^{mast} = 30.0$ mV mostly due to acoustics.

Since a direct beat note between master and reference laser cannot be created due to their large frequency gap, the performance of the transfer cavity is estimated in the following way. With a measured sub-millisecond linewidth $\Delta \nu_{short}^{mast} \approx 100$ kHz of the master laser, the absolute deviations $\Delta V_{lock}^{mast} = 30.0$ mV of the lock signal (inset of figure 6.3(b)) lead to a jitter of the cavity length that can be translated into a frequency mismatch of

$$\Delta \nu_{lock}^{mast} = \frac{\Delta V_{lock}^{mast}}{\Delta V_{short}^{mast}} \Delta \nu_{short}^{mast} \approx 500 \text{ kHz} .$$

Analogously, a frequency mismatch $\Delta \nu_{lock}^{Ref} \approx 100$ kHz can be derived from the corresponding lock signal of the reference laser. This results in a total stability of the frequency offset between signal and reference laser of ≈ 600 kHz. The final discussion of frequency stability will need to take into account the lock performance of the filter cavity setup.

Since transfer and filter cavities cannot be adjusted to exactly the same mirror distance, a tuning parameter for the frequency offset between master and reference lasers is inevitable to ensure maximum transmission of the

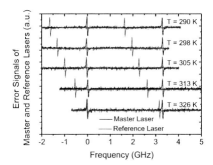

Figure 6.4: Relative frequency shift between master and reference lasers with tuning crystal temperature. Changing the temperature by about 36 K, the reference laser resonance is moved by half an FSR compared to the master.

degenerate DRO output in the filter cavity setup. This tuning is provided by setting the temperature of a birefringent quartz crystal inside the transfer cavity. Using orthogonal polarizations for master and reference laser, the two beams experience ordinary and extraordinary indices of refraction, respectively, which show different dependencies on crystal temperature T. Figure 6.4 depicts the according relative frequency shift between master and reference laser resonances. Spanning a temperature range of ≈ 36 K, a relative frequency shift by half an FSR is achieved which allows easy adjustment for simultaneous resonance of reference laser and DRO output in the filter cavity. Using material parameters provided in [181], the theoretical value for a shift by half an FSR is 31.4 K.

The temperature of the tuning crystal is stable within ± 25 mK. From figure 6.4, a sensitivity of 30 MHz/K for the master-reference frequency offset can be derived. Thus, absolute long-term frequency deviations of ≈ 1.5 MHz are expected which exceeds the short-term lock noise. However, the transmission stays constant within the $\Delta\nu_{FP} = 43$ MHz filter cavity linewidth.

6.2.2 Filter Setup

Fabry-Pérot cavity and etalon are set up inside an isolated box on a temperature stabilized ground plate similar to the transfer setup (photograph in figure B.2). Figure 6.5 provides a schematic of the experimental realization. Since relative frequency stability between reference and master lasers is en-

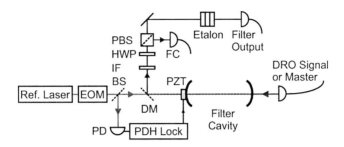

Figure 6.5: Schematic setup of the filter setup. The filter cavity is locked to the reference laser by the PDH technique. The DRO signal laser or master laser counter-propagates through the cavity and is separated by a dichroic mirror (DM), interference filters (IF), and by polarization. It then passes an air-spaced etalon in a second filter stage.

sured by the transfer setup, the reference laser at 852.1 nm can be used for length stabilization of the filter cavity. Using the PDH technique, an EOM modulates the phase of the reference beam. A dispersion signal is derived from the field reflected by the cavity and subsequent lock-in detection at the EOM modulation frequency as mentioned above. Figure 6.6 shows the reflected power from the cavity input mirror as well as the generated dispersion signal for the reference laser when the cavity PZT is scanned over a full FSR. A high coupling efficiency of 0.84 is achieved, and a filter cavity linewidth of 51.2 MHz can be determined from the full FSR scan. The dispersion signal exhibits a signal-to-noise ratio larger than 90 : 1. Following the procedure in the discussion of the transfer setup, the stability of the cavity lock can be derived. With the reference laser linewidth $\Delta\nu_{short}^{Ref} = 100$ kHz, a lock noise $\Delta V_{lock}^{Ref} = 70$ mV and a laser linewidth induced noise $\Delta V_{short}^{Ref} = 20$ mV suggest a filter cavity stability $\Delta\nu_{lock}^{Ref} = 350$ kHz.

The DRO signal field – or the master laser for test measurements – counter-propagates through the filter resonator and is separated from the reference beam by a dichroic mirror (DM). Residual reference photons are attenuated by additional 10 nm wide interference filters (IF). Orthogonal polarization of DRO signal and reference laser allows additional filtering by a PBS that is also used to monitor the master laser transmission spectrum at an additional fiber coupler (FC).

The master laser transmission through the filter cavity is depicted in figure 6.7(a). It reaches a peak value of 0.91, and a filter cavity linewidth $\Delta\nu_{FP}^{exp} \approx 50$ MHz has been measured. The slight deviation from the expected

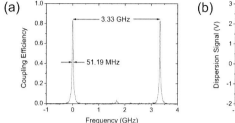 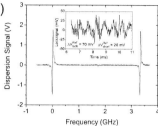

Figure 6.6: (a) Reflected reference laser power from the filter cavity input mirror. A coupling efficiency of 0.84 is measured. (b) Dispersion signal for the reference laser created by the PDH technique. For a discussion of the lock stability, see text.

value of $\Delta\nu_{FP}^{theo} = 43$ MHz is attributed to the inaccuracy of the mirror reflectivities by about 0.004.

Figure 6.1 showed the insufficient filter performance of the Fabry-Pérot cavity due to the large width of the DRO phase-matching envelope. Thus, a second filter stage was proposed that can be realized by a subsequent etalon. The etalon is built in air-spaced configuration in order to maintain high-quality coatings and roughness for the reflecting surfaces. The air slit has a length of 3 mm which corresponds to a free spectral range $\mathrm{FSR}_{Et} = 50.0$ GHz. Temperature stabilization of the etalon within ± 150 mK ensures excellent long-term stability that keeps the transmission above 0.95 of the peak value. A PZT serves as the etalon spacer and allows a variation of the air gap up to 3 μm.

The etalon transmission at a PZT scan over one FSR_{Et} is provided in figure 6.7(b). A peak transmission of around 0.88 has been achieved; scaling the abscissa by the free spectral range, a linewidth of $\Delta\nu_{Et}^{exp} \approx 2.7$ GHz could be derived. The corresponding finesse $F_{Et}^{exp} = 18.5$ does not agree with the desired surface reflectivity of 0.9 ($F_{Et}^{theo} = 29.4$) and is probably caused by degradation due to surface roughness or lack of surface parallelism. Still, it is sufficient for successful filtering of the DRO signal field.

By accounting for the dichroic mirror, interference filters, and fiber coupling, the overall peak transmission through the filter setup under locked conditions can be determined to 0.08. Mainly the interference filters ($T_{IF} = 0.75$ each) and the low fiber coupling efficiency behind the etalon ($T_{FC} = 0.22$) contribute to the attenuation. With the large filter cavity linewidth of 50 MHz, the reduction of this peak transmission due to the lock noise of around 1 MHz

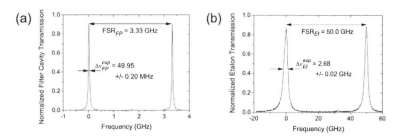

Figure 6.7: (a) Filter cavity transmission spectrum. The PZT scan shows only the fundamental transversal cavity mode with a coupling efficiency of 0.91 and a linewidth $\Delta\nu_{FP} \approx 50$ MHz. (b) Etalon transmission spectrum. A peak transmission of 0.88 could be achieved at a linewidth of $\Delta\nu_{Et}^{exp} \approx$ 2.7 GHz.

and tuning crystal drifts of < 1.5 MHz can be neglected. Figure 6.8 provides a long-term measurement of the filter setup transmission for the master laser over 3 h. The output power stays above 0.8 of the maximum transmission value. The combination of temperature instabilities of tuning crystal (0.05 K) and etalon (0.3 K) as well as lock noise, however, explain fluctuations of only ~ 0.05. Additionally, no correlation between temperature instability and transmission is obvious. The residual deviations may rather be caused by long-term length drifts of the etalon PZT.

6.3   Single-Mode Operation of the DRO

After this characterization using the master laser, the filter setup is applied to the DRO signal field. Fiber coupling provides this quantum field in a TEM_{00} mode, and the linewidth < 3 MHz of its individual longitudinal modes appears as a δ-distribution compared to the linewidths of the filter stages. For this measurement, the DRO was pumped to provide a signal photon rate of $\kappa_{MM}^{exp} \sim 446$ kHz at the filter cavity. Accounting for the filter setup transmission of 0.08 and $N = 45$ effective DRO cavity modes, the theoretical single-mode count rate is estimated as $\kappa_{SM}^{theo} = 1.0$ kHz. This prediction agrees very well with the actually measured count rate of $\kappa_{SM}^{exp} = 1.0 \pm 0.04$ kHz and presents a first indication of single-mode operation. A dark count rate of 0.4 kHz due to residual reference laser light and electronic noise has been subtracted. Final proof, however, can be provided by measuring the signal-idler intensity cross-correlations. While a periodic structure – oscillating at the

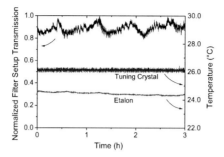

Figure 6.8: Long-term measurement of the transmission through the filter setup for the master laser. The transmission is normalized to the setup transmission of 0.08. Relative power deviations of this peak transmission are < 0.2 over several hours. The temperature instabilities of tuning crystal and etalon, however, are too small to explain these fluctuations. Drifts are therefore attributed to drifts of the etalon PZT.

FSR_S frequency of the DRO resonator – is expected for multi-mode emission (equation 5.2, figure 5.10), single-mode output yields a simple exponential decay.

Experimental data are depicted in figure 6.9. Figure 6.9(a) shows the intensity correlation function between the single-mode signal and multi-mode idler field. The exponential decay reveals a single-mode photon linewidth of FWHM $= 2.84 \pm 0.05$ MHz which coincides with previous multi-mode measurements. There is no evidence of an oscillation at $\text{FSR}_S^{exp} = 1.45$ GHz which is verified by the fast Fourier transform (FFT) in figure 6.9(b). For comparison, the intensity cross-correlations are measured at the same count rate without single-mode filtering. Results are provided in figure 6.9(c), and the corresponding FFT (figure 6.9) clearly shows a peak at FSR_S^{exp}.

6.4 Discussion

In chapter 5, the cavity-enhanced generation of single photons from SPDC could be demonstrated. This result itself provides an important resource for applications in quantum information processing. Quantum information can be encoded onto these narrow-band single-photon states by means of time-bin encoding as described in chapter 7, and fundamental studies of single-photon non-locality or arbitrary shaping of photon wave packets become possible. The multi-mode character of the DRO emission, however, is

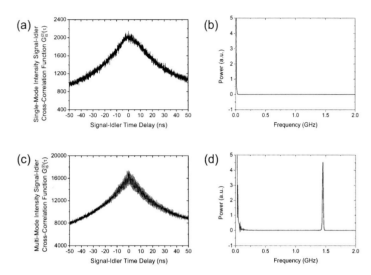

Figure 6.9: (a) Signal-idler cross-correlation function with a single-mode signal field. The function shows a single exponential decay with $\gamma_{S/I} \approx$ 4.44 MHz. (b) Corresponding Fourier spectrum without multi-mode interference. (c) Multi-mode single-idler cross-correlation function for reference. (d) Here, the Fourier spectrum shows a strong periodicity at $\text{FSR}_S = 1.45$ GHz.

not suitable for proposed experiments requiring narrow-band single photons in a single longitudinal resonator mode. Namely, the coupling effects to a single atomic resonance will be overwhelmed by sideband modes which lead to bad signal-to-noise ratio. Thus, single-mode filtering of the DRO signal output has been demonstrated in the present chapter. Single-mode emission is thereby indicated by the achieved count rate of ~ 1 kHz at a DRO pump power of 1 mW and could finally be proven by intensity cross-correlations between signal and idler fields. The usually multi-peaked decaying structure of the correlation function, that announces the presence of multiple longitudinal cavity modes, transforms into a single exponential decay. The single-mode decay constant of $\gamma_{S/I} = 4.44 \pm 0.08$ MHz coincides with previous multi-modes measurements and yields a single-mode single-photon linewidth of only FWHM $= 2.84 \pm 0.05$ MHz. These are excellent conditions for demonstrations of atom-photon interactions on the single-photon level, and pre-studies towards an experimental realization using coherent input fields will be subject of chapter 8.

Chapter 7

Time-Bin Encoding of Narrow-Band Single Photons

"Secrecy is the soul of all great designs"
– Charles Caleb Colton

Narrow-band single photons, which have been generated by cavity-enhanced SPDC, *a priori* carry quantum information since they are emitted into a well-defined polarization mode. Polarization is the generic degree of freedom of a single photon in order to encode quantum information, and various schemes have successfully been demonstrated [126, 182, 183], especially for cryptographic tasks. Unfortunately, polarization encoding may cause serious problems. First, polarization states are not suited for reliable long-distance transmission in optical fibers and require additional techniques for fiber length stabilization. Residual birefringence cannot be avoided even by careful polarization adjustment along the fiber core's principal axes and results in a degradation of the incident state. Second, photon storage of polarization-encoded single photons requires the use of more complicated atomic level systems beyond the Λ scheme. Often cited systems are the tripod [184] or the M configuration [185], but both suffer from additional sources of decoherence and an implementation in Zeeman sublevels for the realization by an actual atomic species. The latter complicates the separation of macroscopic pump and single-photon field.

In order to overcome the obstacles of long-range fiber transmission for quantum information, the concept of time-bin encoding has been developed [47,

105

49]. Using the coherent superposition of two time windows, or *time bins*, for the temporal localization of a single photon, efficient and reliable transmission of qubit states over long distances can be ensured. The explanation of this technique and its realization for narrow-band and thus far-extended wave packets will be discussed in this chapter.

7.1 Foundations of Time-Bin Encoding

As an alternative to polarization encoding, quantum information may be encoded in the global phase of a single photon. In order to prepare and read these states, interferometric techniques are used. If a single photon impinges on the input port of an unbalanced interferometer, its output state is a coherent superposition of both interferometer paths. This output represents an element of a two-dimensional Hilbert space, just as a polarization-encoded state. For a $50 : 50$ input beam splitter, the state can be visualized on the equator of the Bloch sphere with the azimuthal angle ϕ_A represented by the relative phase between the two interferometer modes. One obtains an output state in the superposition

$$|\psi\rangle = \frac{1}{\sqrt{2}} \left(|1, 0\rangle + e^{i\phi_A} |0, 1\rangle \right)$$

where the basis vector $|1, 0\rangle$ characterizes the first and the basis vector $|0, 1\rangle$ stands for the second time bin, corresponding to the short and the long interferometer path, respectively (figure 7.1). Thereby, the spacing δt between the bins must exceed the photon coherence time in order to distinguish reliably between the two basis states.

The readout can be performed equivalently by a second interferometric setup

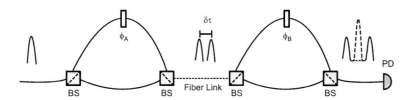

Figure 7.1: Scheme of time-bin encoding using two interferometers. Depending on the relative phase $\phi_B - \phi_A$, constructive or destructive interference of the central peak can be achieved.

displaying a phase difference ϕ_B. The output state behind the second interferometer describes the superposition of three different arrival times

$$|\psi'\rangle = \frac{1}{2} \left(|1,0,0\rangle + e^{i\phi_A}|0,1,0\rangle + e^{i\phi_B}|0,1,0\rangle + e^{i(\phi_A + \phi_B)}|0,0,1\rangle \right) \quad (7.1)$$

where the two central summands account for the indistinguishable paths

$$|0\rangle_L = |\mathrm{short}_A, \mathrm{long}_B\rangle \quad \text{and} \quad |1\rangle_L = |\mathrm{long}_A, \mathrm{short}_B\rangle\,.$$

From equation 7.1, the probability to detect the photon in the central peak is derived as

$$P = |\langle 0,1,0|\psi'\rangle|^2 = \frac{1}{2}\left[1 + \cos\left(\phi_B - \phi_A\right)\right]\,. \quad (7.2)$$

Thus, constructive or destructive interference is determined by the relative phase $\phi_B - \phi_A$. In order to achieve a high visibility, the variation of the arm length should be kept constant within a fraction of the photon's wavelength. This can be achieved by active stabilization of the relative path difference.

Application to narrow-band single photons: For the implementation of time-bin encoding with narrow-band photons, interferometers in a Michelson configuration will be used in this experiment. While time-bin encoding is usually applied to short pulses of picosecond length, narrow-band single photons of about 10 MHz width – as generated by the DRO for efficient atom-photon coupling – represent wave packets with an extension of a few 100 ns which increases the effort for path length stabilization. For 10 MHz spectral width, the separation of single-photon wave packets in the short and long arms requires a path length difference of at least 100 ns. Choosing path length differences of 100 m length in each of the long arms, a time delay of ~ 500 ns is predicted. The wavelength of the DRO signal photons is tuned to

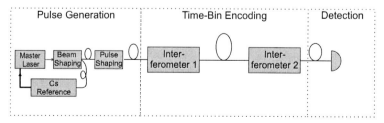

Figure 7.2: Principal building blocks for time-bin encoding of narrow-band photons. For details see text.

894.3 nm which fixes the requirements for relative path length stabilization. The phases ϕ_A and ϕ_B need to be set with an accuracy that allows reliable switching between constructive and destructive interference, as well as corresponding superpositions in the central time bin. This is achieved by an EOM in the short arm of each interferometer. Path length stability is guaranteed by both passive temperature control and active stabilization on a reference beam at 852 nm via a PZT attached to the end mirror of the short arm.

7.2 Experimental Design

An overview of the experimental realization is depicted in figure 7.2, and a photograph can be found in B.4. For test purposes, coherent pulses of variable temporal length can be used as interferometer inputs instead of the DRO signal photons. All individual blocks of the setup are connected by optical fibers in order to provide modularity of the different components. The schemes for pulse generation and the explicit setup of the interferometers will be subject of the following description.

7.2.1 Generation of Short Pulses

In order to characterize the interferometers, a reliable source of macroscopic short pulses is indispensable. Pulses with a width of around 100 ns – corresponding to DRO signal photon wave packets of $\sim$ 10 MHz width – are generated as shown in figure 7.3. The incident laser power P_{in} is modulated by an EOM[1] in double-pass configuration and a PBS. The EOM and its HF

_____
[1]New Focus, NFO-4102-M

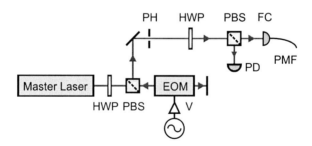

Figure 7.3: Setup for the generation of short test pulses. Amplitude modulation is accomplished by the combination of a PBS and an EOM in double-pass configuration.

driver sit inside a copper box to shield the lab environment against electronic noise and can be adjusted by a five-axes tilt aligner for optimal operation. Depending on the driving EOM voltage V, the intensity at the deflecting output of the PBS follows

$$P_{out} = P_{in} \sin^2\left(\frac{V}{V_\pi} \cdot \frac{\pi}{2} + \phi_0\right)$$

where $V_\pi \approx 280$ V is the half-wave voltage and ϕ_0 the offset phase at $V = 0$. Passing the EOM twice reduces the half-wave voltage to $V_\pi \approx 140$ V in order to switch the polarization from horizontal to vertical and to obtain maximum output into the outbound mode. A pinhole (PH) filters small contaminations by higher TEM modes behind the spatially slightly inhomogeneous EOM crystal. The setup then reaches an extinction ratio of 1 : 50. In general, the driving voltage can be programmed for arbitrary waveform generation with < 25 ns timing accuracy and is amplified by home-made electronics. A photo diode (PD) monitors the pulse shape which simulates single-photon wave packets by Gaussian pulses. These pulses are coupled into the first interferometer via a polarization-maintaining fiber (PMF).

7.2.2   Interferometer Setup

The experimental setup of the Michelson interferometers is shown in figure 7.4. The solid line follows the signal photon field that is split at the central 50 : 50 BS. The necessary path difference is realized by double-passing

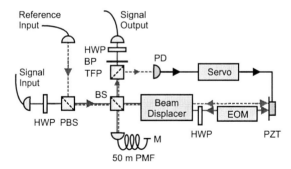

Figure 7.4: Schematic of the interferometric setup. An unbalanced interferometer is realized via 50 m of a double-passed polarization-maintaining fiber. A reference beam travels parallel to the signal field in order to allow for long-term stabilization of the path length difference. For details see text.

a 50 m long PMF with highly reflecting end facet (M) resulting in a delay of about 500 ns. The fiber is wrapped around a massive copper block for temperature stabilization within ± 25 mK, corresponding to an absolute expansion of ± 0.75 μm. In order to stabilize the interferometer's path difference, the signal field is superimposed by a macroscopic reference beam at 852 nm of ~ 100 μW power (dashed line). Signal and reference frequencies differ by a wavelength interval that allows for easy separation by commercial band-pass filters (BP).

Since the reference phase needs to be kept stable when the signal phase is changed by the EOM in the short arm, signal and reference show orthogonal polarization for separation inside a calcite beam displacer. An HWP in front of the EOM sets the signal polarization ensuring only a global phase shift. The short arm end mirror is mounted to a PZT for length stabilization. Signal and reference fields are separated by a thin-film polarizer (TFP) which attenuates spurious reference polarization in the transmitted beam to one part in 10^4. The residual attenuation of the reference beam by eight orders of magnitude for a reliable signal-to-noise ratio in the signal mode is achieved by band-pass filters. The signal output is coupled into a PM fiber.

7.2.3   Stabilization Scheme

The aim of the macroscopic reference beam is the stabilization of the path difference within a fraction of the signal photon wavelength without destroying the signal field by detection. A scheme similar to the stabilization of the filter cavity behind the DRO source is applied which actually uses the same reference laser. The reference laser frequency is stabilized with respect to the master laser via the transfer setup described in section 6.2. A constant frequency offset between reference and master laser – and thus signal photons in the case of degenerate down-conversion in the DRO cavity – yields a fixed phase between the interference fringes for signal and reference fields at the interferometer outputs. Thus, once the relative path length is stabilized to the reference laser, it stays constant for the signal field, as well.

The actual lock of the relative path length to the reference beam is established by monitoring its interference on the photo diode PD. The slope of the interference signal with its nearly linear dependance around the lock point is a direct measure for length fluctuations. It is fed into a servo to provide control signals for both the PZT (figure 7.4) with a control bandwidth around 3 kHz and the temperature stabilization setpoint for the 50 m long fiber in order to compensate for slow drifts using an additional integrator stage. A long-term measurement of the lock performance can be found in subsection 7.3.3.

Beside active length stabilization, special care has been taken to minimize thermal, acoustic, and mechanical noise. Each interferometer is set up inside an isolated box on top of a ground plate that is temperature stabilized within ± 25 mK. All critical components like beam displacers and delay fibers are additionally temperature stabilized. Optical mounts are home-made mini-components for $1/2$ inch optics in order to ensure high mechanical eigenfrequencies. With these precautions, long-term locks up to one hour could be achieved.

7.3 Characterization and Measurements

7.3.1 Pulse Generation

The EOM used for pulse generation is optimized for best extinction ratios by applying a rectangular driving signal at low frequency to monitor the output power by a DC-coupled photo diode. In operation, a voltage for the EOM is generated by an arbitrary waveform DS345 function generator in order to create a Gaussian output pulse of FWHM ≈ 100 ns. Accounting for the transfer function of the EOM setup (equation 7.2.1), this voltage must follow $V \propto \arcsin\left(\sqrt{4\log 2 \exp\left((t-t_0)^2/\mathrm{FWHM}^2\right)}\right)$. Voltage train and output power behind the pinhole are provided in figure 7.5 which proves the reliable generation of pulses with a width of FWHM $= 100.4 \pm 0.2$ ns. The extinction ratio has been optimized via tilt alignment of the EOM and the pulse voltage amplitude.

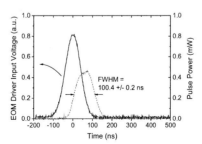

Figure 7.5: Trains of the EOM voltage (solid line) and the measured output light pulse (dashed line). Taking into account the transfer function of the EOM setup, a light pulse of Gaussian shape with 100 ns width is generated.

7.3.2 Signal Mode Visibility

The figure of merit for interferometer performance is its visibility V in the signal mode. This visibility determines the degree of constructive or destructive interference in the central time bin corresponding to the two indistinguishable paths. In case of the reference beam, a high visibility ensures an increased signal-to-noise ratio in the error signal for reliable lock operation.

If the PZT at the short arm end mirror is linearly scanned and a cw input beam is used, the output power follows a sinusoidal shape. The ± 10 V output voltage of the servo controller yields a PZT expansion of 3 μm – corresponding to about four fringes – which can be increased by an intermediary amplifier. Figure 7.6 shows the interference of the signal beam for both interferometers. Visibilities $V_1^{Sign} = 0.87$ and $V_2^{Sign} = 0.92$ can be derived for interferometer 1 and 2, respectively. The non-unity result is mainly attributed to a mode-mismatch in the short arm where the signal mode passes an EOM crystal with slightly tilted facets and residual polarization rotation inside the delay fiber in the long arm. The interference trains of the reference beam will be provided in section 7.3.3 where interferometer stability is discussed.

7.3.3 Interferometer Stabilization

As introduced in section 7.2.3, the interferometer path difference is stabilized via a PZT and temperature control of the delay fiber. In order to obtain an error signal, the interference of the reference beam is monitored on a photo diode, and its slope is used via a servo to provide control of the relative path

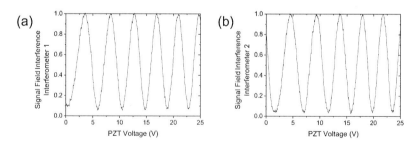

Figure 7.6: Interference of the signal mode in (a) interferometer 1 and (b) interferometer 2. Visibilities of $V_1^{Sign} = 0.87$ and $V_2^{Sign} = 0.92$ could be measured.

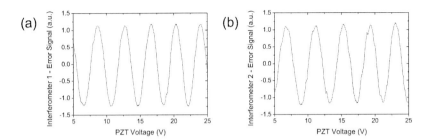

Figure 7.7: Error signals for path length stabilization generated by interference of the reference mode in (a) interferometer 1 and (b) interferometer 2. The signal-to-noise ratios are 22 : 1 and 14 : 1, respectively.

length up to a bandwidth of a few kHz. The interference of the reference laser in both interferometers is displayed in figure 7.7 as a function of their corresponding PZT voltage. Fitting a sine, signal-to-noise ratios of 22 : 1 and 14 : 1 are determined for interferometer 1 and 2, respectively, allowing for reliable stabilization beyond $\lambda/10$. The main limitation of the signal-to-noise ratio is a spurious 50 Hz oscillation visible in both graphs.

Long-term stabilization of the path length difference over 30 min is demonstrated in figure 7.8 for the setup of interferometer 1. The control loops via both PZT and delay fiber temperature are deployed. While the fiber tem-

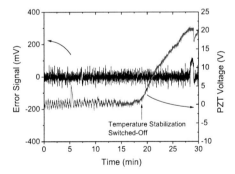

Figure 7.8: Long-term measurement of the interferometer path length stabilization. The error signal is kept stable for nearly 30 min since the lock time was extended by an additional slow control loop via the delay fiber temperature setpoint.

113

perature stabilization is active, the PZT voltage stays close to 0 V. Slight oscillations due to a too large proportional gain in the fiber temperature stabilization loop could not easily be avoided since temperature deviations of a few 10 mK – corresponding to a temperature setpoint change of only 1 mV – result in huge fiber length variations already. These periodic changes are on a slow time scale, however, and can be compensated by the faster PZT control. When the fiber stabilization is switched off, the PZT voltage takes non-zero values. Using the slow control path, however, the PZT voltage stays within its accessible range for an additional 20 min.

7.3.4 Phase Setting for the Signal Mode

With stabilized interferometer path difference, reliable phase setting for the signal mode and switching between constructive and destructive interference will be demonstrated by the following measurements. Generating an incident pulse of 100 ns width as shown in section 7.3.1, the interferometer output displays a double-peak structure, corresponding to transmission through the short and long interferometer arms, respectively (figure 7.9). The time spacing between the peak maxima $\delta t^{exp} = 489.5 \pm 1$ ns matches the path difference of $\Delta s \approx 100.4$ m or $\delta t^{theo} = \Delta s/c \approx 488.3$ ns with an index of refraction $n = 1.46$ for fiber core quartz at 900 nm. This value exceeds the pulse width of 100 ns by far, as required by the time-bin scheme. The pulse shapes themselves stay nearly constant in width with an additional broadening of 4 and 9 ns for the short and long arms, respectively.

The transmission through two consecutive interferometers displays the tem-

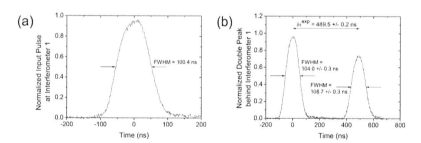

Figure 7.9: (a) Input pulse with a width of 100 ns for interferometer 1 generated by the double-passed EOM. (b) Double peak behind interferometer 1 corresponding to transmission through its short and long arms. The bin interval δt^{exp} matches the optical path length difference $n \Delta s$.

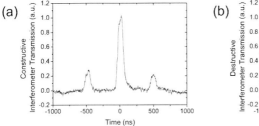 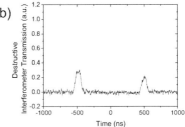

Figure 7.10: Adjusting the relative phase by the EOMs in the short interferometer arms, (a) constructive and (b) destructive interference in the central time bin is observed, corresponding to the states of the time-bin basis. The relative phase has been measured to be long-term stable by locking both interferometers to a common reference beam.

poral characteristics of equation 7.1 and paves the way for the application of the interferometers to time-bin encoding. A corresponding measurement is provided in figure 7.10. If the phase difference between the two interferometers is set to $\phi_B - \phi_A = 2m\,\pi$ with $m \in \mathbb{Z}$, constructive interference is obtained according to equation 7.2 (figure 7.10(a)). Correspondingly, a phase difference $\phi_B - \phi_A = (2m-1)\,\pi$ amounts to a destructive output as displayed in figure 7.10(b). The quality of these two basis states can be characterized by the visibility between constructive and destructive interference in the central bin. The time-bin interval is chosen as δt^{exp}, and gated detection is performed within the central time window only. From the measurements in figure 7.10, a visibility close to unity is derived.

7.4 Discussion

The prerequisites for successful time-bin encoding were defined in section 7.1 as time-bin separation larger then the signal's coherence time, long-term stabilization of the length difference between indistinguishable paths, and reliable phase setting for the time-bin basis states. All these key features have been proven by the presented measurements. Assuming a 10 MHz spectral width of single-photon wave packets, delay fibers produced temporal bins of ≈ 490 ns separation with negligible wave packet distortion. In case of narrower wave packets, this time interval can easily be extended to larger values since fiber absorption does not set a limit to the interferometer transmission.

By applying fast and slow control paths, both interferometer arm length differences were locked to a common reference beam, and long-term stability of the path length differences over hours could be proven. In a last step, the EOMs in the short interferometer arms were used to adjust the relative phase $\phi_B - \phi_A$, and constructive and destructive interference in the central time bin could be observed with nearly perfect visibility. In summary, the properties cited in this chapter show the applicability of time-bin encoding to narrow-band single photons as produced by cavity-enhanced SPDC and pave the way for storage of time-bin encoded quantum information in atomic quantum memories.

Chapter 8

Towards Coupling of Single Photons to Atomic Transitions

> *"Light thinks it travels faster than anything*
> *but it is wrong.*
> *No matter how fast light travels,*
> *it finds the darkness has always got there first,*
> *and is waiting for it."*
>
> – Terry Pratchett

The ultra-narrow linewidth of single photons generated by cavity-enhanced SPDC makes them ideal candidates for efficient coupling to atomic ensembles. This coupling to atomic resonances opens new paths for the study of atom-photon interactions on the single-photon level, but also attracts interest due to its applicability to single-photon storage and quantum memories based on the effect of electromagnetically induced transparency (EIT). Thus, a detailed study of EIT parameters with coherent light fields is indispensable before more complex experimental tasks including single-photon states can be approached.

8.1 EIT

It has been observed in three-level systems that the coupling of two involved levels by a strong coupling field suppresses real and imaginary parts of the linear susceptibility on a second participating transition which can be monitored by a weak probe field. First theoretical and experimental studies in sodium vapor were demonstrated by Alzetta et al. [186] and Whitley and Stroud [187, 188] who named the effect "coherent population trapping". Later, this effect was termed electromagnetically induced transparency (EIT) [189] and extensive studies followed [54, 55, 190, 191]. Since the formation of coherent superpositions is highly limited by decoherence, EIT can be studied most easily in gases where dephasing rates are much lower compared to solids. EIT has been observed in Λ, V, and ladder systems. Experiments around lasing without inversion have primarily been conducted in the V configuration [192–194] while the Λ system is preferred for photon storage requiring two metastable states [195, 196].

Theoretical Background

The goal of this chapter is the study of the EIT effect in hot cesium vapor as a tool for single-photon storage. Thus, a corresponding level scheme in Λ configuration will be considered as depicted in figure 8.1 containing the two hyperfine levels of the $6^2S_{1/2}$ ground state and the $F' = 4$ state of the $6^2P_{1/2}$ fine structure level (see also figure A.1). The two ground state levels $|b\rangle$ and $|c\rangle$ are coupled to the common excited level $|a\rangle$ by coupling and probe fields via dipole-allowed transitions.

Optical Bloch equations: In the dipole approximation, the coupling strengths of coupling and probe fields are characterized by Rabi frequencies

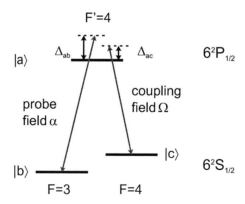

Figure 8.1: Definitions of energy levels and laser transitions for the observation of EIT on the cesium D1 line.

Ω and α, respectively. The Rabi frequency for a transition between two magnetic sublevels $|F_i, m_{F_i}\rangle \rightarrow |F_j, m_{F_j}\rangle$ is defined as

$$\nu_{Rabi} = \frac{1}{2\pi} \frac{2\left\langle F_j, m_{F_j}\left|e\,\vec{r}\right|F_i, m_{F_i}\right\rangle \vec{E}_q^{(+)}}{\hbar}$$

where $\vec{E}_q^{(+)}$ is the positive frequency part of the driving field with polarization q in the spherical basis. For the transitions between two hyperfine states in the considered scheme (figure 8.1), an effective dipole moment can be introduced, and the fast oscillating part of $\vec{E}_q^{(+)}$ will be separated. The driving field is then expressed by $2\left|\vec{E}_q^{(+)}\right| \rightarrow E_0 = \sqrt{I/(2c\epsilon_0)}$ with intensity I.

For the presented pre-studies, the levels will be coupled by coherent laser fields; thus, the light can be treated classically, and the system Hamiltonian reads

$$H = \hbar\omega_a|a\rangle\langle a| + \hbar\omega_b|b\rangle\langle b| + \hbar\omega_c|c\rangle\langle c| - \hbar\alpha e^{2\pi i\nu_p t}|a\rangle\langle b| - \hbar\alpha e^{2\pi i\nu_p t}|a\rangle\langle c| + \text{h.a.}$$

with energies $\hbar\omega_x$ ($x \in \{a,b,c\}$) of the atomic levels and the laser frequencies ν_c and ν_p of coupling and probe field, respectively. Time evolution is described by the Liouville equation

$$\dot{\rho} = -\frac{i}{\hbar}[H, \rho]$$

which simplifies to a set of linear equations in case of a stationary situation. In order to account for incoherent processes like damping or dephasing, one makes use of the master equation formalism, including the coupling to a reservoir in thermal equilibrium [197]. Then, the optical Bloch equations for the described system follow

$$\dot{\rho}_{ab} = -\left(i\omega_{ab} + \frac{\gamma_a + \gamma_{bc}}{2}\right)\rho_{ab} + i\alpha e^{-2\pi i\nu_p t}\left(\rho_{bb} - \rho_{aa}\right) + i\Omega e^{-2\pi i\nu_c t}\rho_{cb}$$

$$\dot{\rho}_{ac} = -\left(i\omega_{ac} + \frac{\gamma_a + \gamma_{bc}}{2}\right)\rho_{ac} + i\Omega e^{-2\pi i\nu_c t}\left(\rho_{cc} - \rho_{aa}\right) + i\alpha e^{-2\pi i\nu_p t}\rho_{bc}$$

$$\dot{\rho}_{cb} = -\left(i\omega_{cb} + \gamma_{bc} + \gamma_{deph}\right)\rho_{cb} + i\Omega^* e^{2\pi i\nu_c t}\rho_{ab} - i\alpha e^{-2\pi i\nu_p t}\rho_{ca}$$

$$\dot{\rho}_{aa} = -\gamma_a\rho_{aa} + i\alpha e^{-2\pi i\nu_p t}\rho_{ba} - i\alpha^* e^{2\pi i\nu_p t}\rho_{ab} + i\Omega e^{-2\pi i\nu_c t}\rho_{ca} - i\Omega^* e^{2\pi i\nu_c t}\rho_{ac}$$

$$\dot{\rho}_{bb} = -\gamma_{bc}\rho_{bb} + \gamma_{bc}\rho_{cc} + \gamma_{ab}\rho_{aa} - i\alpha e^{-2\pi i\nu_p t}\rho_{ba} + i\alpha^* e^{2\pi i\nu_p t}\rho_{ab}$$

$$\dot{\rho}_{cc} = -\gamma_{bc}\rho_{cc} + \gamma_{bc}\rho_{bb} + \gamma_{ac}\rho_{aa} - i\Omega e^{-2\pi i\nu_c t}\rho_{ca} + i\Omega^* e^{2\pi i\nu_c t}\rho_{ac}$$

Here, $\gamma_a = \gamma_{ab} + \gamma_{ac}$ considers spontaneous emission from the excited state $|a\rangle$, and atomic transition frequencies $\omega_{ij} := \omega_i - \omega_j$ with $i, j \in \{a,b,c\}$. The rate γ_{bc} describes population exchange between the two ground states due to the movement of atoms in and out of the interaction zone which is assumed symmetric ($\gamma_{bc} = \gamma_{cb}$). All pure dephasing mechanisms between $|b\rangle$ and $|c\rangle$ without population exchange, like laser linewidths and magnetic stray fields, are integrated into γ_{deph}.

Applying the rotating wave approximation [197], a stationary solution of the optical Bloch equations is obtained by assuming $\dot{\rho} \equiv 0$ and normalization $\mathrm{Tr}\,(\rho) = 1$. Coupling and probe fields are assumed collinear, so the residual Doppler shift $(k_c - k_p)\,v$ is neglected in further calculations. Due to the weak intensity of the probe laser, the problem can be treated perturbatively with $\rho^{(1)} = \rho^{(0)} + \Delta\rho$ where $\rho^{(0)}$ describes the solution for $\alpha = 0$. Then, the explicit expression for the considered coherence $\rho_{ab}^{(1)}$ on the probe laser transition is

$$
\rho_{ab}^{(1)} = -\frac{i\alpha}{\left(\gamma_{bc} + \gamma_{deph} + i\Delta_{bc}\right)\left(\frac{\gamma_a + \gamma_{bc}}{2} + i\Delta_{ab}\right) + \Omega^2} \times
$$

$$
\times \left[\gamma_{bc} + \gamma_{deph} + i\Delta_{bc}\left(\rho_{aa}^{(0)} - \rho_{bb}^{(0)}\right) + \frac{\Omega^2}{\left(\frac{\gamma_a + \gamma_{bc}}{2} - i\Delta_{ac}\right)}\left(\rho_{cc}^{(0)} - \rho_{aa}^{(0)}\right)\right].
$$

$$(8.1)$$

The detunings have been defined according to $\Delta_{ab} := (\omega_a - \omega_b) - \nu_p$, $\Delta_{ac} := (\omega_a - \omega_c) - \nu_c$ and $\Delta_{bc} := \Delta_{ab} - \Delta_{ac}$.

Absorption and Dispersion: Equation 8.1 allows the calculation of the linear susceptibility via

$$
\chi = \eta\,\frac{\rho_{ab}^{(1)}}{\alpha} \qquad \text{with} \qquad \eta = \frac{3}{8\pi^2}N_{Cs}\gamma_a\lambda_{ba}^3 . \tag{8.2}
$$

N_{Cs} denotes the cesium density, and $\lambda_{ba} = 2\pi c/(\omega_a - \omega_b)$ the probe laser wavelength on resonance. From the susceptibility, the absorption coefficient α_{EIT} and the group velocity v_g can be determined and will be derived in the following.

The complex index of refraction is defined by $\tilde{n} = n - i\,\kappa$ with extinction coefficient κ. Using $\chi = \tilde{n}^2 - 1$, one finds $\mathrm{Im}\,[\chi] = \mathrm{Im}\,[n^2 - \kappa^2 - 2in\kappa - 1]$. Since $n \cong 1$ in dilute vapor, $\mathrm{Im}\,[\chi] \cong -2\kappa$ follows. This leads to [157]

$$
\alpha_{EIT} = \frac{2\omega\kappa}{c} = -\frac{2\pi}{\lambda_{ba}}\,\mathrm{Im}\,[\chi]
$$

and the transmission coefficient for a gas cell of length L_{Cs}

$$
T = \exp\left(-\alpha_{EIT}L_{Cs}\right).
$$

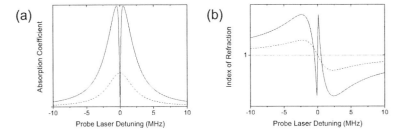

Figure 8.2: (a) Absorption and (b) index of refraction on the probe laser transition. The absorption shows the typical feature of high transmission on resonance while the index of refraction exhibits a steep slope as required for low group velocities. The dashed lines give the dependencies without coupling laser for reference.

According to [195], dispersive properties like the index of refraction $n \sim \mathrm{Re}\,[1 + \chi]^1$ or the group velocity

$$v_g = \mathrm{Re} \left[\frac{c - \nu \frac{\partial(1+\chi)}{\partial k}}{(1 + \chi) + \nu \frac{\partial(1+\chi)}{\partial \nu}} \right]$$

can be calculated. While the first summand in v_g accounts for frequency dispersion, the second is attributed to spatial dispersion. The experimentally accessible parameter is the pulse delay inside the cell

$$\tau = L_{Cs} \left(\frac{1}{v_g} - \frac{1}{c} \right) . \tag{8.3}$$

The dependance of absorption coefficient and index of refraction on probe laser detuning is depicted in figure 8.2. Figure 8.2(a) shows the absorption coefficient with corresponding parameters of the considered cesium transitions according to figure 8.1. The solid line gives the probe absorption for a coupling laser Rabi frequency $\Omega = 0.5$ MHz while the dashed line provides the dependance without coupling laser for reference. The transparency window is visible only in the presence of the coupling laser. Figure 8.2(b) exhibits the corresponding index of refraction with a steep slope if the coupling laser is applied (solid line). Again, the dashed line provides the situation without coupling laser, showing anomalous dispersion around zero detuning.

---

[1]Due to the definition of the Rabi frequencies in frequency, not angular frequency space, no additional factor 2π is required.

In order to obtain realistic fits for experimental data, Doppler broadening needs to be accounted for in equation 8.2. Atomic classes of different velocities $\vec{v}$ contribute according to their relative motion with respect to the wave vectors $\vec{k_p} \parallel \vec{k_c}$ of the laser fields. The Doppler broadened expression then follows a convolution of equation 8.2 with the atomic velocity distribution

$$\chi = \eta \int d\left(\vec{k}\vec{v}\right) f\left(\vec{k}\vec{v}\right) \frac{\rho_{ab}^{(1)}\left(\vec{k}\vec{v}\right)}{\alpha} . \tag{8.4}$$

For analytic results, the Maxwell-Boltzmann distribution is approximated by a Lorentzian [198, 199]

$$f\left(\vec{k}\vec{v}\right) = \frac{\Delta w_D/\left(2\pi\right)}{\left(\Delta w_D/2\right)^2 + \left(\vec{k}\vec{v}\right)^2} .$$

This allows the evaluation of the integral via the residue theorem. Δw_D is the Doppler line width according to the cesium vapor temperature T_{Cs}, and detunings in equation 8.1 must be substituted according to

$$\Delta_{ac} \to \Delta_{ac} + \vec{k}\vec{v}/\left(2\pi\right) , \quad \Delta_{ab} \to \Delta_{ab} + \vec{k}\vec{v}/\left(2\pi\right) , \quad \text{and} \quad \Delta_{bc} \to \Delta_{bc} .$$

All relevant parameters for a comparison between this theory and experimental results on the cesium D1 line will be cited from [200]. The coupling laser will be tuned to resonance, so $\Delta_{ac} = 0$ follows. The decoherence rate is determined by time-of-flight broadening due to atoms moving in and out of the coupling laser beam and can be calculated by

$$\gamma_{bc} = \frac{\sqrt{2k_B T_{Cs}/M_{Cs}}}{4\pi w} .$$

Here, M_{Cs} denotes the mass of a cesium atom and w the mean beam waist inside the cell. The most intriguing modification of the transparency window is the scaling of its width with $\Delta\nu_{EIT}^{Dopp} \sim \Omega^2/\Delta w_D$ in the case of Doppler broadening instead of $\Delta\nu_{EIT} \sim \Omega^2/\gamma_a$ for a gas at absolute zero [201].

8.2 Experimental Setup

The observation of EIT in hot cesium vapor on the transitions displayed in figure 8.1 requires two laser sources, each tuned to their cesium hyperfine transition with appropriately shaped pulses and a cell that is temperature stabilized, magnetically shielded, and contains the atomic species. The following paragraphs will give an introduction to the main parts of the setup.

8.2.1 Frequency Stabilization of Coupling and Probe Laser

As coupling and probe laser sources, two extended-cavity diode lasers[2] are coupled into single-mode fibers in order to provide high-purity TEM_{00} profiles. The lasers need to be frequency stabilized to interact efficiently with the respective transitions $6^2S_{1/2}(F = 3) \rightarrow 6^2P_{1/2}(F' = 4)$ for the probe and $6^2S_{1/2}(F = 4) \rightarrow 6^2P_{1/2}(F' = 4)$ for the coupling laser. For the coupling laser, this is performed by FMS (see section 4.4) in order to allow for stabilization on individual hyperfine transitions by Doppler-free spectroscopy. EIT absorption profiles can be obtained by scanning the probe laser across the $6^2S_{1/2}(F = 3) \rightarrow 6^2P_{1/2}(F' = 4)$ resonance while measurements of pulse delays, caused by the steep dispersion slope, require a fixed stabilization to two-photon resonance. Therefore, the probe laser is frequency offset locked with respect to the stabilized coupling laser as sketched in figure 8.3 [202, 203]. This provides the necessary offset of ~ 9.2 GHz between the ground state hyperfine levels.

Both laser beams are superimposed on a PBS and projected into a linear diagonal polarization for heterodyning on a fast photo diode. The beat note at around $\nu_{Synth} \approx 9.2$ GHz is mixed with a synthesizer frequency to yield an RF signal in an electronically processable frequency range which is compared to a 30 MHz local oscillator (LO). For phase differences beyond 2π, the phase

---

[2]Toptica Photonics, DL100

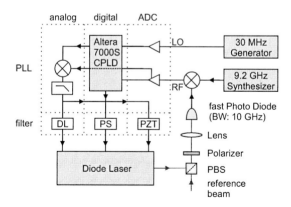

Figure 8.3: Scheme for the analog/digital hybrid system for frequency offset lock between coupling and probe laser. For details see text.

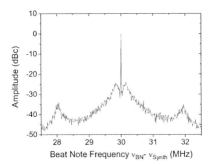

Figure 8.4: Beat note spectrum showing the performance of the frequency offset lock with a spectrum analyzer resolution of 3 kHz. The contribution of the central peak is 0.165.

difference is determined digitally via a complex programmable logic device (CPLD, Altera, EPM7064) to generate a corresponding voltage for the laser lock (digital phase detector). Within a phase difference of 2π, an analog phase detector is deployed consisting of a mixer with a subsequent low-pass filter. In analog or digital mode, the error voltage is fed into two filter paths. The first makes use of the laser's extended-cavity PZT to account for low frequency noise < 1 Hz and long-term drifts. The second controls the laser current via a modulation input of the laser power source (PS) with a band-width up to $\sim$ 80 kHz. For higher frequency laser noise, the analog control provides an additional third filter path (DL) feeding the laser diode directly via a bias-T which reaches a bandwidth of $\sim$ 2 MHz.

The system performance is depicted in figure 8.4. In principle, this system is designed for phase-locking two diode lasers, but the observation of EIT requires a stable frequency offset lock only. Up to 0.165 of the total beat note power are measured in the central peak of the spectrum with a spectrum analyzer at 3 kHz resolution. The bumps at ± 2 MHz relative to the carrier frequency can be contributed to a spurious resonance in the DL filter path and may be shifted to a higher frequency by a redesign of the lead circuit in that filter path.

8.2.2 EIT Setup

The total setup for observation of the EIT effect in cesium vapor is depicted in figure 8.5. The offset of probe and coupling laser frequencies is locked, and the

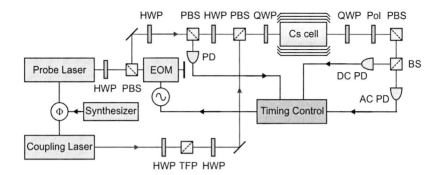

Figure 8.5: Overal scheme of the optical setup for the generation of EIT in hot cesium vapor. For details see text.

probe laser can be varied in lock by changing the synthesizer frequency ν_{Synth}. In order to observe slow light, the probe laser power must be modulated. This is achieved by the setup described in detail in section 7.3.1 using a double-passed EOM and a PBS. The probe laser power may be modulated either sinusoidally or following a Gaussian shape by driving the EOM with an arbitrary waveform generator which allows a time resolution of ~ 25 ns. The polarizations of coupling and probe lasers are carefully set to vertical and horizontal, respectively, with > 0.9999 purity and superimposed at a PBS. To perform EIT studies in the circular basis, an additional QWP in front of the cesium cell is added. A second QWP, or a corresponding HWP for experiments with linear polarizations inside the cell, as well as a polarizer (Pol) are set to deflect the probe beam at a final PBS. The probe power is then monitored by a DC coupled photo diode to record EIT spectra and a fast AC coupled detector for pulse measurements. The photo diode in front of the combining PBS monitors power fluctuations of the probe laser for reference.

The EIT cell itself is filled with ^{133}Cs specified to be 0.9999 monoisotopic. It has a length of 4 cm and a diameter of 25 mm. The number of interacting cesium atoms at a given temperature can be calculated via the vapor pressure curve [200] and the beam waists. Beam waists of coupling and probe lasers inside the cell are $w_c = 1.4$ mm and $w_p = 1.1$ mm, respectively. With these parameters, the corresponding Rabi frequencies can be calculated. In order to minimize losses and the formation of parasitic etalons, the end facets of the gas cell are AR coated for 894 nm. Since Earth's magnetic field and stray fields induce level shifts and decoherence between the two considered hyperfine ground states, the gas cell is surrounded by three layers of μ-metal (figure B.1). The considered shield design attenuates DC magnetic fields by a factor of $\sim 2 \times 10^5$ [204] which means a negligible residual line shift of ~ 1 Hz due to Earth's magnetic field. A heating foil is glued to the cell surface using a heat conductive silica film to allow for temperature setting between 25 and 80 °C. The long-term temperature stability has been measured to ± 10 mK over 10 h.

8.3 Observation of EIT

For pre-studies, EIT spectra at room temperature were observed without the frequency offset lock. The probe laser was scanned across the $6^2S_{1/2}(F = 3) \rightarrow 6^2P_{1/2}(F' = 4)$ resonance by the extended-cavity PZT of the probe laser while the coupling laser stayed locked to $6^2S_{1/2}(F = 4) \rightarrow 6^2P_{1/2}(F' = 4)$. Results at a cesium temperature $T_{Cs} = 39$ °C are depicted in figure 8.6. The

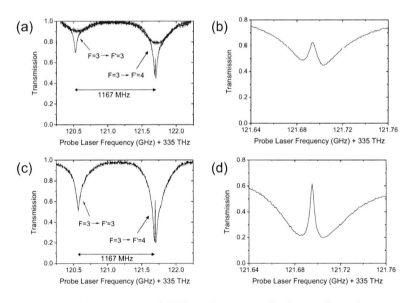

Figure 8.6: Measurement of EIT on the cesium D1 line without frequency offset lock. (a) Absorption and EIT spectra before optimization and probe laser transmission for reference with (b) providing a zoom into the EIT peak. (c) Corresponding measurement after optimization of relevant beam parameters. (d) Zoom into the improved EIT peak.

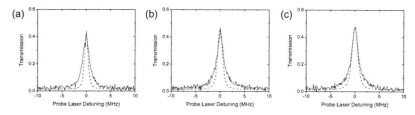

Figure 8.7: EIT peaks at varying coupling laser power P_c. The dashed lines are theoretical fits according to equation 8.4 depending on γ_{deph} as the only free parameter. (a) $P_c = 2$ mW, $\gamma_{deph} = 200$ kHz, (b) $P_c = 3$ mW, $\gamma_{deph} = 235$ kHz, and (c) $P_c = 4$ mW, $\gamma_{deph} = 280$ kHz.

two upper figures 8.6(a) and (b) provide a first visual proof of EIT with the characteristic transparency window on the proper hyperfine line at a coupling laser power of 4.1 mW. However, parameters of the two beams had not been optimized showing an EIT transparency of only 0.32. By adjusting probe and coupling laser polarizations and powers, beam profile matching, and checking on the collinearity of probe and coupling lasers, the EIT transparency could be improved to 0.52 as shown in figures 8.6(c) and (d) at a coupling laser power of 2 mW. The probe laser power did not exceed a few 10 μW.

Detailed studies of the dependance of transparency and EIT window width on temperature and coupling laser power, however, require the application of the frequency offset lock. Thereby, the probe laser is tuned by ± 500 MHz relative to its resonance on the $6^2S_{1/2}(F = 3) \rightarrow 6^2P_{1/2}(F' = 4)$ hyperfine transition by changing the synthesizer frequency. The frequency resolution is set to 50 kHz within the central ± 10 MHz region and to 10 MHz outside. Zooms into the transparency window of typical EIT spectra are provided in figure 8.7 for a coupling laser power varying between 2 and 4 mW. The probe laser power was set to 80 μW at a cell temperature of 39 °C where transmission outside the transparency window is nearly negligible. The transparency shows a value around 0.45. The dashed lines provide the theoretically expected transmission according to the model presented in subsection 8.1 with a nearly constant dephasing rate $\gamma_{deph} \approx 240$ kHz as the only free parameter.

A second experimental parameter with high impact on the EIT features is the temperature of the cesium cell. Measurements have been performed for temperatures between 25 and 60 °C at powers of 4 mW and 80 μW for coupling and probe lasers, respectively (figure 8.8). The graphs focus on the central EIT transparency window. With increasing temperature and Doppler broadening, the width of the transparency window drops according to equa-

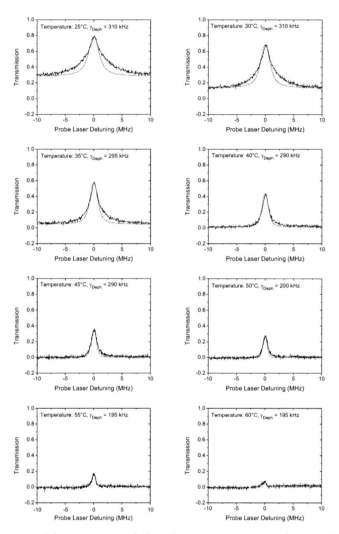

Figure 8.8: Measurements of the EIT transparency window at different cesium cell temperatures. The increase of opacity and the simultaneous decrease of transparency window width with rising temperature follows the theoretical predictions (solid lines).

tion 8.4 (solid lines). Correspondingly, the opacity increases due to higher cesium density, and an optical depth > 3 is reached at a cell temperature of 35 °C already. The population exchange rate γ_{bc} accounts for the temperature dependant atom velocity, scaling with $\sqrt{T_{Cs}}$, and only the dephasing rate γ_{deph} is used to fit the experimental data. The decrease of γ_{deph} with rising temperature could not be explained by the current theory and will be discussed in section 8.5.

8.4   Slow Light in Cesium

The primary purpose of the EIT effect in this experimental context is the slowing and eventually storage of light. Thus, the potential of the setup for group velocity reduction inside the cesium cell is evaluated in detail and compared the theoretical predictions according to equation 8.3. In order to measure light delay, the probe laser is locked to the $6^2S_{1/2}(F=3) \rightarrow 6^2P_{1/2}(F'=4)$ resonance. Its intensity is modulated by the EOM depicted in figure 8.5. Figure 8.9(a) provides the result for sinusoidal modulation. Experimental parameters were a coupling laser power $P_c = 2$ mW and a cesium temperature $T_{Cs} = 30$ °C. According to the measurements in figure 8.8, a dephasing rate $\gamma_{deph} = 310$ kHz is expected. Theory then predicts a delay of $\tau^{theo} = 186$ ns which is close to the experimental value of $\tau^{exp} \approx 195$ ns. For comparison, a change of the dephasing rate by just 30 kHz produces an error of ~ 10 ns. Figure 8.9(b) shows the delay of a Gaussian shaped probe laser

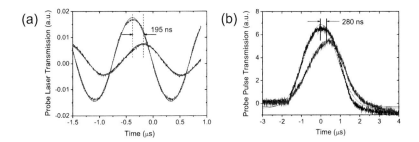

Figure 8.9: (a) Delay of a sinusoidal modulation of the probe laser power. A value of 195 ns is achieved for $T_{Cs} = 30$ °C at $P_c = 2$ mW. (b) Delay of a Gaussian shaped probe laser pulse by 280 ns at $P_c = 0.5$ mW and $T_{Cs} = 30$ °C. The solid lines show fit curves according to equation 8.3 using the corresponding dephasing rate $\gamma_{deph} = 310$ kHz.

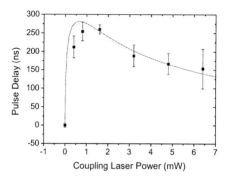

Figure 8.10: Dependance of pulse delay on coupling laser power. The maximum delay is achieved around 1 mW. The solid line gives a theoretical fit using a dephasing rate $\gamma_{deph} = 310$ kHz.

pulse (FWHM = 2 μs, 80 μW peak power) at the same cell temperature and $P_c = 0.5$ mW. The measured delay of $\tau^{exp} \approx 280$ ns matches the theoretical prediction of $\tau^{theo} = 262$ ns which assumes $\gamma_{deph} = 310$ kHz. Again, the discrepancy can be explained by a < 30 kHz mismatch between the dephasing rates obtained from figure 8.8 and in the current measurement. For a comprehensive characterization, pulse delays (FWHM = 2 μs, 80 μW peak power) were recorded dependant on coupling laser power with $T_{Cs} \approx 31$ °C. Each data point in figure 8.10 is an average of four measurements with error bars representing absolute uncertainties. The delay is maximum around a coupling power of 1 mW.

8.5 Discussion

In this chapter, EIT on the cesium D1 line has been studied with special attention to the achievable group delay. Parameters of the transparency window have been measured, and excellent agreement with a theoretical approach based on a single free parameter could be obtained. This single parameter γ_{deph} accounts for pure dephasing between the two hyperfine ground states and takes a value around $200 - 300$ kHz in our setup. An EIT window with up to 0.60 transparency and $1 - 2$ MHz width was achieved, limited by dephasing due to laser frequency noise. A slight mismatch between experiment and theory is visible in the wings of the transparency window. Therefore, the influence of the probe laser intensity on the EIT spectra has been examined in more detail. With parameters $P_c = 4$ mW and $T_{Cs} = 30$ °C,

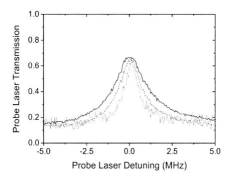

Figure 8.11: EIT transparency windows at probe powers of $P_p = 34$ μW (dotted line), $P_p = 66$ μW (dashed line), and $P_p = 120$ μW (solid line). The width of the window increases with probe power.

the results in figure 8.11 actually prove this discrepancy to originate from the influence of the finite probe laser power that is not completely covered by the low excitation regime assumed in the theory. With increasing probe power, the transparency window broadens similar to the saturation effect of a strong coupling laser. Unfortunately, the detectors in use did not allow a further reduction of the probe laser power.

Beyond the observation of transparency effects, group velocity reduction in the cesium cell has been studied. Pulse delays up to 280 ns were achieved which corresponds to a reduction of the speed of light by a factor of ~ 2000 for a 4 cm long gas cell. This delay crucially depends on the coupling laser power and takes its maximum around 1 mW. At lower coupling laser power, population cannot be pumped effectively into the $6^2S_{1/2}(F = 3)$ state. At higher power, saturation broadens the transparency window and leads to shallower dispersion. Both effects deteriorate the achievable group delay.

Chapter 9

Conclusions and Outlook

Within this thesis, various approaches are demonstrated that show the applicability of single photons to tasks in the field of quantum information processing. As an outline in the introductory chapter, figure 1.2 provided the scheme for the experimental study of atom-photon interactions on the single-photon level. Along this road, this thesis describes the generation of narrow-band single photons using cavity-enhanced SPDC. In a first step, only the signal field is resonated in a pump-enhanced single-resonant OPO far below threshold, and a spectral photon width of 62 MHz has been achieved. For the first time, the stabilization of the OPO resonator to the pump beam enables long-term operation of cavity-enhanced SPDC over hours. A new theoretical description of this system from first principles has been developed that allows the extraction of principal parameters of this source by intensity cross-correlation measurements.

Beyond the scope of a single-resonant device, work was started on an OPO resonating both signal and idler fields, taking advantage of overall count rate enhancement and the prospect of even narrower single-photon linewidth. A setup could be realized that exhibits triple resonance of signal and idler as well as pump fields in an OPO cavity driven far below threshold, again allowing for long-term stabilization. The linewidth of this setup reaches an outstanding value of 2.7 MHz, and its brightness of 14000 counts/s per mW pump power and MHz signal bandwidth surpasses all previous realizations by at least two orders of magnitude. Characterizing signal-idler cross-correlation measurements show excellent agreement with predicted dependencies that have been derived from an extended version of the single-resonant theory. Exploiting the concept of heralded single-photon generation, antibunching in the signal mode has been demonstrated reaching $g_c^{(2)}(0) = 0.012 \pm 0.005$ at a heralding rate $R_H = 5$ kHz. From this data, the suppression of higher photon number contributions could be derived, yielding a 100-fold improvement compared to a Poissonian source of the same brightness. The heralding efficiency of this source reaches $P(1) = 0.55 \pm 0.01$. All these features make this double-resonant OPO far below threshold a remarkable tool for the study of fundamental questions of quantum mechanics, e.g., experiments about single-photon non-locality which is easily accessible due to extended photon wave packets of a few 100 ns. Exploration of atom-photon interaction, however, requires the signal field to be emitted into a single longitudinal OPO mode. This issue is tackled by a two-stage filter setup that selectively transmits a single frequency mode out of a comb of 45, attenuating residual modes by more than two orders of magnitude. Measuring intensity cross-correlations between the idler and the filtered signal field, the single-mode character could be proven. With a determined single-photon linewidth of 2.8 MHz in a single longitudinal mode and built-in stabilization to the cesium D1 line, this source is perfectly suited for efficient coupling to an atomic transition.

In order to prepare these narrow-band single photons for applications in quantum computing or cryptography, interferometric setups were developed for quantum information encoding by the time-bin scheme. Accounting for the corresponding extended wave packets of narrow-band photons, unbalanced interferometers with 100 m arm length difference have actively been stabilized by a reference laser. This reference laser is indirectly locked to the OPO output frequencies, thereby avoiding to destroy signal or idler states by detection. Together with an experimental setup that additionally focuses on passive stability against mechanic, thermal, and acoustic noise, reliable time-bin encoding could be achieved over hours using laser pulses for test purposes. Imprinting quantum bits in the time-bin degree of freedom of narrow-band single-photons paves the way for quantum information experiments towards atom-photon coupling. The efficient coupling between atomic states and single photons from a purely photonic source – contrary to photons generated from atoms themselves – has not been demonstrated so far. Thus, light-matter interaction in cesium has been studied extensively in order to prepare such experiments. Especially coherent excitation in three-level systems is of great interest since it allows the observation of effects like EIT and photon storage. EIT has been measured in a cell containing cesium hot vapor, and group velocity reduction by a factor of more than 2000 could be obtained. Thus, principal building blocks of figure 1.2 could be realized. Combined with a single-mode single-photon bandwidth of < 3 MHz generated by cavity-enhanced SPDC, these are promising results for the realization of a atom-photon interface and further of a quantum memory.

The next steps along this road will certainly concern improvements of the EIT setups. In order to reduce the dephasing rate between the two ground state levels, a phase lock system between coupling and probe laser has been set up

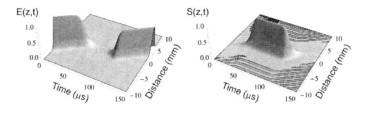

Figure 9.1: Simulation of dynamic EIT. The photonic excitation E vanishes with decreasing coupling laser power after ~ 40 μs. It is transformed into a stationary spin coherence S between the hyperfine ground states of the EIT Λ scheme. The photon can be retrieved by turning the coupling laser on, again.

with state-of-the-art phase noise variance $\sigma_\phi^2 = 0.02$ rad^2 [205]. This should allow for higher transmission and wider transparency windows. The actual storage of photons also requires dynamic EIT schemes where an adiabatic coupling laser switch-off transfers the photonic excitation onto a long-living spin state [56, 196, 206]. A simulation of the photonic and spin contributions is depicted in figure 9.1. Once the coupling laser is switched off after ~ 40 μs, the photonic excitation vanishes. During this adiabatic process, the photon becomes a stationary coherence between the two meta-stable ground states of the EIT Λ scheme. Turning the coupling laser on, again, revives the photonic excitation. The necessary intensity modulation of the coupling laser is already implemented in the experimental setup and will allow for much longer storage times.

The time-bin setup has reached a stage that reliable encoding of quantum information on single photons can be performed. The transmission of a secret key by the BB84 protocol [7] will be used to demonstrate this capability in a proof-of-principle experiment. A program for automatic timing control of the BB84 algorithm has been implemented that synchronizes laser pulse generation, EOM settings, and gated readout for quantum key distribution. Measured pulse trains are depicted in figure 9.2. After randomly setting the phase information by Alice's and Bob's EOMs, a light pulse is generated, and the central maximum of the triple peak behind both interferometers is selected by APD gating. In order to realize quantum cryptography with a narrow-band single-photon source, the heralding idler field of the DRO will be used as the trigger source for the transmission protocol.

The performance of the DRO allows direct application to further experiments around single-photon non-locality. Arbitrary shaping of narrow-band single-

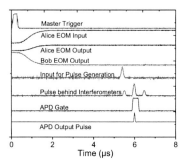

Figure 9.2: Measured trains of control signals for a synchronized BB84 implementation.

photon wave packets has been performed by [207]. However, their use of a single-photon source based on Raman transitions in an atomic ensemble restricts important experimental parameters due to natural constants of the selected atomic species, including wave packet length and central frequency. Here, the use of cavity-enhanced SPDC offers a substantial advantage of versatility. The obtained linewidth of < 3 MHz and corresponding extension > 50 ns or ~ 16 m allows the straightforward implementation by an EOM for pulse shaping.

As mentioned, the capability of the DRO to precisely control single-photon emission can also be used in multiple schemes for the realization of atom-photon interactions on the single-photon level. Two explicit ideas – connecting single photons from cavity-enhanced SPDC and a source based on stimulated Raman transitions – will be sketched in the following.

Single-photon indistinguishability: As a basic problem of experimental quantum information processing at the present stage, many physical systems exist separately, but in general, qubits cannot be exchanged between them. This limits the available number of qubits in many implementations and avoids the combination of advantages certain physical realizations provide. For optical implementations of quantum information processing, different schemes of single-photon generation have special features suited for certain tasks. Single photons generated by cavity-enhanced SPDC and Raman scattering in atomic ensembles may be connected in a single experiment in order to show the indistinguishability of single photons from two independent physical systems. Raman scattering in atomic ensembles presents a single-photon source at room temperature with unique features and a nat-

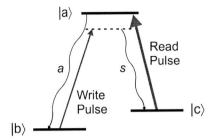

Figure 9.3: The detection of a Stokes photon s heralds an excitation in state $|c\rangle$ after applying a non-resonant write pulse. This excitation is reconverted into an anti-Stokes photon a by a resonant read pulse.

ural link to atomic transitions [208, 209]. Assuming the three-level system in figure 9.3, an off-resonant write laser drives a Raman transition at very low probability. At appropriate conditions, only a single excitation will be stored in ground state $|c\rangle$, and its existence is heralded by a Stokes photon s on the $|a\rangle \to |c\rangle$ transition. For an ensemble of atoms, this excitation will be uniformly distributed among the atoms in a Dicke state which makes this state very robust against atom losses. A resonant read laser pulse is used to reconvert this excitation into a single anti-Stokes photon a. This source thus provides all the advantages of a stationary system compared to the photonic approach with flying qubits.

As a first step towards connecting these two sources, one should show the indistinguishability of their emitted photons. The wave packets emitted by SPDC can be made identical to photons emitted by the Raman scheme using cavity-enhancement as described before. The central wavelength is set via the phase-matching condition of the nonlinear crystal while its bandwidth can be adjusted by parameters of the surrounding cavity. The indistinguishability may then proven by a Hong-Ou-Mandel experiment; due to their bosonic character, indistinguishable photon pairs will always leave into a same output port of a $50 : 50$ beam splitter.

SPDC-mediated entanglement generation: An interesting application of these two matched single-photon sources is the generation of entanglement between two distant atomic ensembles L and R, mediated by a cavity-enhanced SPDC source. In the original scheme by Duan, Lukin, Cirac, and Zoller for entanglement generation between two distant atomic ensembles (DLCZ, figure 9.4) [16], entanglement is created by applying a non-resonant write pulse to each ensemble. Then a click at exactly one detector behind

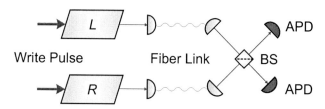

Figure 9.4: Entanglement generation between atomic ensembles according to the DLCZ proposal. Entanglement is heralded by the detection of a single photon at only one of the two APD modules.

a $50 : 50$ beam splitter configuration, heralds entanglement between the two ensembles in the form

$$|\psi\rangle = |0\rangle_L |1\rangle_R + \exp(i\phi)|1\rangle_L |0\rangle_R = (S_L + \exp(i\phi)S_R) |0\rangle_L |0\rangle_R$$

where S_L and S_R represent the spin-flip operators for each ensemble and ϕ accounts for different path lengths between the ensembles and the beam splitter. This state represents two of the four Bell states for $\phi = 0$ and $\phi = \pi$. In the scheme proposed here, an SPDC source and two atomic ensembles are used as shown in figure 9.5. Signal and idler outputs are assumed frequency degenerate and tuned to an atomic transition. Next, one lets interfere the scattered Stokes photons from the ensembles with one down-converted photon each. The click at exactly one detector of each detector pair heralds the generation of exactly one excitation in each ensemble. The simultaneous arrival of all four photons at the two beam splitters can be ensured by synchronization of the write beams and by pulsed pumping of the down-conversion source. The detection of exactly one photon in each detector assembly heralds either the generation of a signal-idler pair or exactly one excitation in either ensemble. This state is also maximally entangled and takes the form

$$\begin{aligned} |\psi\rangle &= |0\rangle_L |0\rangle_R + \exp\left(i(\phi_L + \phi_R)\right) |0\rangle_L |0\rangle_R \\ &= \left(\mathbb{1} + \exp\left(i(\phi_L + \phi_R)S_L S_R\right) |0\rangle_L |0\rangle_R . \end{aligned}$$

It represents exactly the other two missing Bell states. As an additional advantage, the two atomic ensembles can be located at twice the attenuation length of the communication channel while four ensembles are needed for the same distance in the usual scheme. Thus, two atomic ensembles can be replaced by a single cavity-enhanced SPDC source which further reduces resources for the architecture of a quantum network.

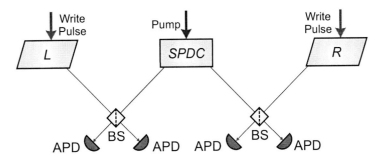

Figure 9.5: Proposed scheme for entanglement generation including two atomic ensembles and a cavity-enhanced SPDC source. The two ensembles L and R can be separated by twice the attenuation length of the communication channel.

These two proposals demonstrate the indistinguishability of photons from different physical systems for the first time. Cavity-enhanced SPDC is also shown to be capable of entanglement creation between atomic ensembles in a new way. Such a scheme would be a novel combination of atomic and purely optical realizations of single-photon sources and outlines the potential of the realized DRO source for a larger quantum network.

Bibliography

[1] A. Einstein, B. Podolsky, and N. Rosen. Can quantum-mechanical description of physical reality be considered complete? *Physical Review*, 47(10):777–780, 1935. doi: 10.1103/physrev.47.777. URL http://link.aps.org/abstract/PR/v47/p777.

[2] A. Aspect, J. Dalibard, and G. Roger. Experimental test of Bell's inequalities using time- varying analyzers. *Physical Review Letters*, 49 (25):1804–1807, 1982. doi: 10.1103/physrevlett.49.1804. URL http://link.aps.org/abstract/PRL/v49/p1804.

[3] G. Weihs, T. Jennewein, C. Simon, H. Weinfurter, and A. Zeilinger. Violation of Bell's inequality under strict Einstein locality conditions. *Physical Review Letters*, 81(23):5039–5043, 1998. doi: 10.1103/physrevlett.81.5039. URL http://link.aps.org/doi/10.1103/PhysRevLett.81.5039.

[4] W. H. Brattain. Demonstration of a bipolar transistor. Technical report, Bell Laboratories, 1947.

[5] T. H. Maiman. Stimulated optical radiation in ruby. *Nature*, 187(4736): 493–494, 1960. doi: 10.1038/187493a0. URL http://dx.doi.org/10.1038/187493a0.

[6] R. Feynman. Simulating physics with computers. *International Journal of Theoretical Physics*, 21(6):467–488, 1982. doi: 10.1007/bf02650179. URL http://dx.doi.org/10.1007/BF02650179.

[7] C. H. Bennett and G. Brassard. Quantum cryptography: Quantum key distribution and coin tossing. In *Proceedings of the IEEE International Conference on Computers, Systems and Signal Processing, Bangalore, Indien*, pages 175–179, 1984.

[8] N. Gisin, G. Ribordy, W. Tittel, and H. Zbinden. Quantum cryptography. *Rev. Mod. Phys.*, 74(1):145–195, 2002. doi: 10.1103/revmodphys.

74.145. URL `http://link.aps.org/doi/10.1103/RevModPhys.74.145`.

[9] L. K. Grover. A fast quantum mechanical algorithm for database search. In *STOC '96: Proceedings of the twenty-eighth annual ACM symposium on Theory of computing*, pages 212–219, New York, NY, USA, 1996. ACM. doi: 10.1145/237814.237866. URL `http://doi.acm.org/10.1145/237814.237866`.

[10] P. W. Shor. Polynomial-time algorithms for prime factorization and discrete logarithms on a quantum computer. *SIAM Journal on Computing*, 26(5):1484–1509, 1997. doi: 10.1137/s0097539795293172. URL `http://link.aip.org/link/?SMJ/26/1484/1`.

[11] E. Knill, R. Laflamme, and G. J. Milburn. A scheme for efficient quantum computation with linear optics. *Nature*, 409(6816):46–52, 2001. doi: 10.1038/35051009. URL `http://dx.doi.org/10.1038/35051009`.

[12] S. Gasparoni, J. Pan, P. Walther, T. Rudolph, and A. Zeilinger. Realization of a photonic controlled-not gate sufficient for quantum computation. *Physical Review Letters*, 93(2):020504, 2004. doi: 10.1103/physrevlett.93.020504. URL `http://link.aps.org/doi/10.1103/PhysRevLett.93.020504`.

[13] N. Kiesel, C. Schmid, U. Weber, G. Tóth, O. Gühne, R. Ursin, and H. Weinfurter. Experimental analysis of a four-qubit photon cluster state. *Physical Review Letters*, 95(21):210502, 2005. doi: 10.1103/physrevlett.95.210502. URL `http://link.aps.org/doi/10.1103/PhysRevLett.95.210502`.

[14] J. I. Cirac, P. Zoller, H. J. Kimble, and H. Mabuchi. Quantum state transfer and entanglement distribution among distant nodes in a quantum network. *Physical Review Letters*, 78(16):3221–3224, 1997. doi: 10.1103/physrevlett.78.3221. URL `http://link.aps.org/doi/10.1103/PhysRevLett.78.3221`.

[15] S. Lloyd, M. S. Shahriar, J. H. Shapiro, and P. R. Hemmer. Long distance, unconditional teleportation of atomic states via complete Bell state measurements. *Physical Review Letters*, 87(16):167903, 2001. doi: 10.1103/physrevlett.87.167903. URL `http://link.aps.org/doi/10.1103/PhysRevLett.87.167903`.

[16] L. Duan, M. D. Lukin, J. I. Cirac, and P. Zoller. Long-distance quantum communication with atomic ensembles and linear optics. *Nature*, 414(6862):413–418, 2001. doi: 10.1038/35106500. URL http://dx.doi.org/10.1038/35106500.

[17] F. Schmidt-Kaler, H. Häffner, M. Riebe, S. Gulde, G. P. T. Lancaster, T. Deuschle, C. Becher, C. F. Roos, J. Eschner, and R. Blatt. Realization of the Cirac-Zoller controlled-not quantum gate. *Nature*, 422(6930):408–411, 2003. doi: 10.1038/nature01494. URL http://dx.doi.org/10.1038/nature01494.

[18] D. Leibfried, B. DeMarco, V. Meyer, D. Lucas, M. Barrett, J. Britton, W. M. Itano, B. Jelenkovic, C. Langer, T. Rosenband, and D. J. Wineland. Experimental demonstration of a robust, high-fidelity geometric two ion-qubit phase gate. *Nature*, 422(6930):412–415, 2003. doi: 10.1038/nature01492. URL http://dx.doi.org/10.1038/nature01492.

[19] T. Wilk, S. C. Webster, A. Kuhn, and G. Rempe. Single-atom single-photon quantum interface. *Science*, 317(5837):488–490, 2007. doi: 10.1126/science.1143835. URL http://www.sciencemag.org/cgi/content/abstract/317/5837/488.

[20] P. Michler, A. Kiraz, C. Becher, W. V. Schönfeld, P. M. Petroff, L. Zhang, E. Hu, and A. Imamoglu. A quantum dot single-photon turnstile device. *Science*, 290(5500):2282–2285, 2000. doi: 10.1126/science.290.5500.2282. URL http://www.sciencemag.org/cgi/content/abstract/290/5500/2282.

[21] J. E. Mooij, T. P. Orlando, L. Levitov, L. Tian, C. H. van der Wal, and S. Lloyd. Josephson persistent-current qubit. *Science*, 285(5430): 1036–1039, 1999. doi: 10.1126/science.285.5430.1036. URL http://www.sciencemag.org/cgi/content/abstract/285/5430/1036.

[22] Y. A. Pashkin, T. Yamamoto, O. Astafiev, Y. Nakamura, D. V. Averin, and J. S. Tsai. Quantum oscillations in two coupled charge qubits. *Nature*, 421(6925):823–826, 2003. doi: 10.1038/nature01365. URL http://dx.doi.org/10.1038/nature01365.

[23] H. J. Briegel, W. Dür, J. I. Cirac, and P. Zoller. Quantum repeaters: The role of imperfect local operations in quantum communication. *Physical Review Letters*, 81(26):5932–5935, 1998. doi:

143

10.1103/physrevlett.81.5932. URL http://link.aps.org/doi/10.1103/PhysRevLett.81.5932.

[24] D. Bouwmeester, J. Pan, K. Mattle, M. Eibl, H. Weinfurter, and A. Zeilinger. Experimental quantum teleportation. *Nature*, 390(6660): 575–579, 1997. doi: 10.1038/37539. URL http://dx.doi.org/10.1038/37539.

[25] E. Moreau, I. Robert, L. Manin, V. Thierry-Mieg, J. Gérard, and I. Abram. Quantum cascade of photons in semiconductor quantum dots. *Physical Review Letters*, 87(18):183601, 2001. doi: 10.1103/physrevlett.87.183601. URL http://link.aps.org/doi/10.1103/PhysRevLett.87.183601.

[26] P. Grangier, B. Sanders, and J. Vuckovic. Focus on single photons on demand. *New Journal of Physics*, 6, 2004. doi: 10.1088/1367-2630/6/1/e04. URL http://stacks.iop.org/1367-2630/6/i=1/a=E04.

[27] A. Kuhn, M. Hennrich, and G. Rempe. Deterministic single-photon source for distributed quantum networking. *Physical Review Letters*, 89(6):067901, 2002. doi: 10.1103/physrevlett.89.067901. URL http://link.aps.org/abstract/PRL/v89/e067901.

[28] M. Keller, B. Lange, K. Hayasaka, W. Lange, and H. Walther. Continuous generation of single photons with controlled waveform in an ion-trap cavity system. *Nature*, 431(7012):1075–1078, 2004. doi: 10.1038/nature02961. URL http://dx.doi.org/10.1038/nature02961.

[29] C. Brunel, B. Lounis, P. Tamarat, and M. Orrit. Triggered source of single photons based on controlled single molecule fluorescence. *Physical Review Letters*, 83(14):2722–2725, 1999. doi: 10.1103/physrevlett.83.2722. URL http://link.aps.org/abstract/PRL/v83/p2722.

[30] B. Lounis and W. E. Moerner. Single photons on demand from a single molecule at room temperature. *Nature*, 407(6803):491–493, 2000. doi: 10.1038/35035032. URL http://dx.doi.org/10.1038/35035032.

[31] P. Michler, A. Imamoglu, M. D. Mason, P. J. Carson, G. F. Strouse, and S. K. Buratto. Quantum correlation among photons from a single quantum dot at room temperature. *Nature*, 406(6799):968–970, 2000. doi: 10.1038/35023100. URL http://dx.doi.org/10.1038/35023100.

[32] M. Nirmal, B. O. Dabbousi, M. G. Bawendi, J. J. Macklin, J. K. Trautman, T. D. Harris, and L. E. Brus. Fluorescence intermittency in single

cadmium selenide nanocrystals. *Nature*, 383(6603):802–804, 1996. doi: 10.1038/383802a0. URL `http://dx.doi.org/10.1038/383802a0`.

[33] P. Frantsuzov, M. Kuno, B. Janko, and R. A. Marcus. Universal emission intermittency in quantum dots, nanorods and nanowires. *Nature Physics*, 4(5):519–522, 2008. doi: 10.1038/nphys1001. URL `http://dx.doi.org/10.1038/nphys1001`.

[34] C. Kurtsiefer, S. Mayer, P. Zarda, and H. Weinfurter. Stable solid-state source of single photons. *Physical Review Letters*, 85(2):290–293, 2000. doi: 10.1103/physrevlett.85.290. URL `http://link.aps.org/abstract/PRL/v85/p290`.

[35] A. Beveratos, S. Kühn, R. Brouri, T. Gacoin, J. Poizat, and P. Grangier. Room temperature stable single-photon source. *The European Physical Journal D - Atomic, Molecular, Optical and Plasma Physics*, 18(2):191–196, 2002. doi: 10.1140/epjd/e20020023. URL `http://dx.doi.org/10.1140/epjd/e20020023`.

[36] Z. Yuan, B. E. Kardynal, R. M. Stevenson, A. J. Shields, C. J. Lobo, K. Cooper, N. S. Beattie, D. A. Ritchie, and M. Pepper. Electrically driven single-photon source. *Science*, 295(5552):102–105, 2002. doi: 10.1126/science.1066790. URL `http://www.sciencemag.org/cgi/content/abstract/295/5552/102`.

[37] A. Badolato, K. Hennessy, M. Atature, J. Dreiser, E. Hu, P. M. Petroff, and A. Imamoglu. Deterministic coupling of single quantum dots to single nanocavity modes. *Science*, 308(5725):1158–1161, 2005. doi: 10.1126/science.1109815. URL `http://www.sciencemag.org/cgi/content/abstract/308/5725/1158`.

[38] D. A. Kleinman. Theory of optical parametric noise. *Physical Review*, 174(3):1027–1041, 1968. doi: 10.1103/physrev.174.1027. URL `http://link.aps.org/abstract/PR/v174/p1027`.

[39] R. L. Byer and S. E. Harris. Power and bandwidth of spontaneous parametric emission. *Physical Review A (Atomic, Molecular, and Optical Physics)*, 168(3):1064–1068, 1968. doi: 10.1103/physrev.168.1064. URL `http://link.aps.org/abstract/PR/v168/p1064`.

[40] L. Mandel and E. Wolf. *Optical coherence and quantum optics*. Cambridge University Press, Cambridge, 1995. ISBN 978-0521417112.

[41] P. G. Kwiat, K. Mattle, H. Weinfurter, A. Zeilinger, A. V. Sergienko, and Y. Shih. New high-intensity source of polarization-entangled photon pairs. *Physical Review Letters*, 75(24):4337–4341, 1995. doi: 10.1103/physrevlett.75.4337. URL http://link.aps.org/abstract/PRL/v75/p4337.

[42] D. C. Burnham and D. L. Weinberg. Observation of simultaneity in parametric production of optical photon pairs. *Physical Review Letters*, 25(2):84–87, 1970. doi: 10.1103/physrevlett.25.84. URL http://link.aps.org/abstract/PRL/v25/p84.

[43] Y. J. Lu and Z. Y. Ou. Optical parametric oscillator far below threshold: Experiment versus theory. *Physical Review A (Atomic, Molecular, and Optical Physics)*, 62(3):033804, 2000. doi: 10.1103/physreva.62.033804. URL http://link.aps.org/abstract/PRA/v62/e033804.

[44] C. E. Kuklewicz, F. N. C. Wong, and J. H. Shapiro. Time-bin-modulated biphotons from cavity-enhanced down-conversion. *Physical Review Letters*, 97(22):223601, 2006. doi: 10.1103/physrevlett.97.223601. URL http://link.aps.org/abstract/PRL/v97/e223601.

[45] J. S. Neergaard-Nielsen, B. M. Nielsen, H. Takahashi, A. I. Vistnes, and E. S. Polzik. High purity bright single photon source. *Optics Express*, 15(13):7940–7949, 2007. doi: 10.1364/oe.15.007940. URL http://www.opticsexpress.org/abstract.cfm?URI=oe-15-13-7940.

[46] X. Bao, Y. Qian, J. Yang, H. Zhang, Z. Chen, T. Yang, and J. Pan. Generation of narrow-band polarization-entangled photon pairs for atomic quantum memories. *Physical Review Letters*, 101(19):190501, 2008. doi: 10.1103/physrevlett.101.190501. URL http://link.aps.org/abstract/PRL/v101/e190501.

[47] A. Muller, T. Herzog, B. Huttner, W. Tittel, H. Zbinden, and N. Gisin. "Plug and play" systems for quantum cryptography. *Applied Physics Letters*, 70(7):793–795, 1997. doi: 10.1063/1.118224. URL http://link.aip.org/link/?APL/70/793/1.

[48] J. Brendel, N. Gisin, W. Tittel, and H. Zbinden. Pulsed energy-time entangled twin-photon source for quantum communication. *Physical Review Letters*, 82(12):2594–2597, 1999. doi: 10.1103/physrevlett.82.2594. URL http://link.aps.org/abstract/PRL/v82/p2594.

[49] G. Ribordy, J. Brendel, J. Gautier, N. Gisin, and H. Zbinden. Long-distance entanglement-based quantum key distribution. *Physical Review A (Atomic, Molecular, and Optical Physics)*, 63(1):012309, 2000. doi: 10.1103/physreva.63.012309. URL http://link.aps.org/ abstract/PRA/v63/e012309.

[50] I. Marcikic, H. de Riedmatten, W. Tittel, V. Scarani, H. Zbinden, and N. Gisin. Time-bin entangled qubits for quantum communication created by femtosecond pulses. *Physical Review A (Atomic, Molecular, and Optical Physics)*, 66(6):062308, 2002. doi: 10.1103/physreva.66. 062308. URL http://link.aps.org/abstract/PRA/v66/e062308.

[51] D. Bouwmeester, A. Ekert, and A. Zeilinger. *The Physics of Quantum Information: Quantum Cryptography, Quantum Teleportation, Quantum Computation.* Springer, Berlin, 2000. ISBN 978-3540667780.

[52] A. Kuzmich, W. P. Bowen, A. D. Boozer, A. Boca, C. Chou, L. Duan, and H. J. Kimble. Generation of nonclassical photon pairs for scalable quantum communication with atomic ensembles. *Nature*, 423(6941): 731–734, 2003. doi: 10.1038/nature01714. URL http://dx.doi.org/ 10.1038/nature01714.

[53] C. H. van der Wal, M. D. Eisaman, A. André, R. L. Walsworth, D. F. Phillips, A. S. Zibrov, and M. D. Lukin. Atomic memory for correlated photon states. *Science*, 301(5630):196–200, 2003. doi: 10.1126/science.1085946. URL http://www.sciencemag.org/cgi/ content/abstract/301/5630/196.

[54] K. Boller, A. Imamoglu, and S. E. Harris. Observation of electromagnetically induced transparency. *Physical Review Letters*, 66 (20):2593–2596, 1991. doi: 10.1103/physrevlett.66.2593. URL http: //link.aps.org/abstract/PRL/v66/p2593.

[55] S. E. Harris. Electromagnetically induced transparency with matched pulses. *Physical Review Letters*, 70(5):552–555, 1993. doi: 10.1103/ physrevlett.70.552. URL http://link.aps.org/abstract/PRL/v70/ p552.

[56] M. Fleischhauer and M. D. Lukin. Dark-state polaritons in electromagnetically induced transparency. *Physical Review Letters*, 84 (22):5094–5097, 2000. doi: 10.1103/physrevlett.84.5094. URL http: //link.aps.org/abstract/PRL/v84/p5094.

[57] M. Bajcsy, A. S. Zibrov, and M. D. Lukin. Stationary pulses of light in an atomic medium. *Nature*, 426(6967):638–641, 2003. doi: 10.1038/ nature02176. URL http://dx.doi.org/10.1038/nature02176.

[58] C. P. Sun, Y. Li, and X.F. Liu. Quasi-spin-wave quantum memories with a dynamical symmetry. *Physical Review Letters*, 91(14):147903, 2003. doi: 10.1103/physrevlett.91.147903. URL http://link.aps. org/abstract/PRL/v91/e147903.

[59] D. Akamatsu, K. Akiba, and M. Kozuma. Electromagnetically induced transparency with squeezed vacuum. *Physical Review Letters*, 92(20): 203602, 2004. doi: 10.1103/physrevlett.92.203602. URL http://link. aps.org/abstract/PRL/v92/e203602.

[60] M. D. Eisaman, A. André, F. Massou, M. Fleischhauer, A. S. Zibrov, and M. D. Lukin. Electromagnetically induced transparency with tunable single-photon pulses. *Nature*, 438(7069):837–841, 2005. doi: 10. 1038/nature04327. URL http://dx.doi.org/10.1038/nature04327.

[61] T. Chanelière, D. N. Matsukevich, S. D. Jenkins, S. Lan, T. A. B. Kennedy, and A. Kuzmich. Storage and retrieval of single photons transmitted between remote quantum memories. *Nature*, 438(7069): 833–836, 2005. doi: 10.1038/nature04315. URL http://dx.doi.org/ 10.1038/nature04315.

[62] D. Gottesman and I. L. Chuang. Demonstrating the viability of universal quantum computation using teleportation and single-qubit operations. *Nature*, 402(6760):390–393, 1999. doi: 10.1038/46503. URL http://dx.doi.org/10.1038/46503.

[63] C. H. Bennett, G. Brassard, C. Crépeau, R. Jozsa, A. Peres, and W. K. Wootters. Teleporting an unknown quantum state via dual classical and Einstein-Podolsky-Rosen channels. *Physical Review Letters*, 70 (13):1895–1899, 1993. doi: 10.1103/physrevlett.70.1895. URL http: //link.aps.org/doi/10.1103/PhysRevLett.70.1895.

[64] B. Zhao, Z. Chen, Y. Chen, J. Schmiedmayer, and J. Pan. Robust creation of entanglement between remote memory qubits. *Physical Review Letters*, 98(24):240502, 2007. doi: 10.1103/physrevlett.98.240502. URL http://link.aps.org/abstract/PRL/v98/e240502.

[65] L. Jiang, J. M. Taylor, and M. D. Lukin. Fast and robust approach to long-distance quantum communication with atomic ensembles. *Physical Review A (Atomic, Molecular, and Optical Physics)*,

76(1):012301, 2007. doi: 10.1103/physreva.76.012301. URL http://link.aps.org/abstract/PRA/v76/e012301.

[66] Z. Chen, B. Zhao, Y. Chen, J. Schmiedmayer, and J. Pan. Fault-tolerant quantum repeater with atomic ensembles and linear optics. *Physical Review A (Atomic, Molecular, and Optical Physics)*, 76(2): 022329, 2007. doi: 10.1103/physreva.76.022329. URL http://link.aps.org/abstract/PRA/v76/e022329.

[67] J. Pan, D. Bouwmeester, H. Weinfurter, and A. Zeilinger. Experimental entanglement swapping: Entangling photons that never interacted. *Physical Review Letters*, 80(18):3891–3894, 1998. doi: 10.1103/physrevlett.80.3891. URL http://link.aps.org/abstract/PRL/v80/p3891.

[68] Z. Zhao, Y. Chen, A. Zhang, T. Yang, H. J. Briegel, and J. Pan. Experimental demonstration of five-photon entanglement and open-destination teleportation. *Nature*, 430(6995):54–58, 2004. doi: 10.1038/nature02643. URL http://dx.doi.org/10.1038/nature02643.

[69] D. N. Matsukevich and A. Kuzmich. Quantum state transfer between matter and light. *Science*, 306(5696):663–666, 2004. doi: 10.1126/science.1103346. URL http://www.sciencemag.org/cgi/content/abstract/306/5696/663.

[70] Y. Chen, S. Chen, Z. Yuan, B. Zhao, C. Chuu, J. Schmiedmayer, and J. Pan. Memory-built-in quantum teleportation with photonic and atomic qubits. *Nature Physics*, 4(2):103–107, 2008. doi: 10.1038/nphys832. URL http://dx.doi.org/10.1038/nphys832.

[71] D. Deutsch. Quantum theory, the Church-Turing principle and the universal quantum computer. *Proceedings of the Royal Society of London. Series A, Mathematical and Physical Sciences (1934-1990)*, 400(1818):97–117, 1985. doi: 10.1098/rspa.1985.0070. URL http://dx.doi.org/10.1098/rspa.1985.0070.

[72] D. Deutsch and R. Jozsa. Rapid solution of problems by quantum computation. *Proceedings of the Royal Society: Mathematical and Physical Sciences (1990-1995)*, 439(1907):553–558, 1992. doi: 10.1098/rspa.1992.0167. URL http://dx.doi.org/10.1098/rspa.1992.0167.

[73] G. I. Taylor. Interference fringes with feeble light. In *Proceedings of the Cambridge Philosophical Society*, volume 15, pages 114–115, 1909.

[74] R. J. Glauber. The quantum theory of optical coherence. *Physical Review*, 130(6):2529–2539, 1963. doi: 10.1103/physrev.130.2529. URL http://link.aps.org/abstract/PR/v130/p2529.

[75] D. F. Walls and G. J. Milburn. *Quantum Optics*. Springer, Berlin, 1994. ISBN 978-3540588313.

[76] Single photon counting module – SPCM-AQR series specifications. Technical report, PerkinElmer, Canada, 2005.

[77] R. Hanbury Brown and R. Q. Twiss. A test of a new type of stellar interferometer on Sirius. *Nature*, 178(4541):1046–1048, 1956. doi: 10.1038/1781046a0. URL http://dx.doi.org/10.1038/1781046a0.

[78] T. Aichele. *Detection and Generation of Non-Classical Light States from Single Quantum Emitters*. PhD thesis, Humboldt-Universität zu Berlin, 2005.

[79] C. B. Murray, D. J. Norris, and M. G. Bawendi. Synthesis and characterization of nearly monodisperse CdE (E = sulfur, selenium, tellurium) semiconductor nanocrystallites. *Journal of the American Chemical Society*, 115(19):8706–8715, 1993. doi: 10.1021/ja00072a025. URL http://pubs.acs.org/doi/abs/10.1021/ja00072a025.

[80] M. A. Hines and P. Guyot-Sionnest. Synthesis and characterization of strongly luminescing ZnS-capped CdSe nanocrystals. *Journal of Physical Chemistry*, 100(2):468–471, 1996. doi: 10.1021/jp9530562. URL http://pubs.acs.org/doi/abs/10.1021/jp9530562.

[81] D. Bimberg, M. Grundmann, and N. Ledentsov. *Quantum Dot Heterostructures*. Wiley VCH, New York, 1998. ISBN 978-0471973881.

[82] M. Scheffler and R. Zimmermann, editors. *Proceedings of 23rd International Conference on The Physics of Semiconductors, Berlin, Germany, 1996*. World Scientific, Singapore. ISBN 981-0227779.

[83] E. Dekel, D. Gershoni, E. Ehrenfreund, J. M. Garcia, and P. M. Petroff. Carrier-carrier correlations in an optically excited single semiconductor quantum dot. *Physical Review B (Condensed Matter and Materials Physics)*, 61(16):11009–11020, 2000. doi: 10.1103/physrevb.61.11009. URL http://link.aps.org/doi/10.1103/PhysRevB.61.11009.

[84] M. Bayer, G. Ortner, O. Stern, A. Kuther, A. A. Gorbunov, A. Forchel, P. Hawrylak, S. Fafard, K. Hinzer, T. L. Reinecke, S. N. Walck, J. P.

Reithmaier, F. Klopf, and F. Schäfer. Fine structure of neutral and charged excitons in self-assembled In(Ga)As/(Al)GaAs quantum dots. *Physical Review B (Condensed Matter and Materials Physics)*, 65(19): 195315, 2002. doi: 10.1103/physrevb.65.195315. URL `http://link.aps.org/doi/10.1103/PhysRevB.65.195315`.

[85] L. Bányai and S. W. Koch, editors. *Semiconductor Quantum Dots*, volume 2 of *World Scientific Series on Atomic, Molecular, and Optical Proper*. World Scientific, Singapore, 1993. ISBN 978-9810213909.

[86] I. N. Stranski and L. Krastanov. Zur Theorie der orientierten Ausscheidung von Ionenkristallen aufeinander. *Sitzungsberichte der Akademie der Wissenschaften in Wien, Abt. IIb*, 146:797–810, 1937. doi: 10. 1007/bf01798103. URL `http://dx.doi.org/10.1007/BF01798103`.

[87] L. Goldstein, F. Glas, J. Y. Marzin, M. N. Charasse, and G. L. Roux. Growth by molecular beam epitaxy and characterization of InAs/GaAs strained-layer superlattices. *Applied Physics Letters*, 47(10):1099–1101, 1985. doi: 10.1063/1.96342. URL `http://link.aip.org/link/?APL/47/1099/1`.

[88] Y.-W. Mo, D. E. Savage, B. S. Swartzentruber, and M. G. Lagally. Kinetic pathway in Stranski-Krastanov growth of Ge on Si(001). *Physical Review Letters*, 65(8):1020–1023, 1990. doi: 10.1103/physrevlett.65. 1020. URL `http://link.aps.org/doi/10.1103/PhysRevLett.65.1020`.

[89] N. Carlsson, W. Seifert, A. Petersson, P. Castrillo, M. Pistol, and L. Samuelson. Study of the two-dimensional–three-dimensional growth mode transition in metalorganic vapor phase epitaxy of GaInP/InP quantum-sized structures. *Applied Physics Letters*, 65(24):3093–3095, 1994. doi: 10.1063/1.112447. URL `http://link.aip.org/link/?APL/65/3093/1`.

[90] M. Ikezawa, Y. Masumoto, T. Takagahara, and S. V. Nair. Biexciton and triexciton states in quantum dots in the weak confinement regime. *Physical Review Letters*, 79(18):3522–3525, 1997. doi: 10.1103/physrevlett.79.3522. URL `http://link.aps.org/doi/10.1103/PhysRevLett.79.3522`.

[91] J. Persson, T. Aichele, V. Zwiller, L. Samuelson, and O. Benson. Three-photon cascade from single self-assembled InP quantum dots. *Physical Review B (Condensed Matter and Materials Physics)*, 69(23):233314,

2004. doi: 10.1103/physrevb.69.233314. URL http://link.aps.org/abstract/PRB/v69/e233314.

[92] M. Born and E. Wolf. *Principles of Optics.* Pergamon Press, Oxford, 1993. ISBN 0-08-026481-6.

[93] T. Wilson and C. J. R. Sheppard. *Theory and Practice of Scanning Optical Microscopy.* Academic Press, London, 1984. ISBN 0-12-757760-2.

[94] C. Kurtsiefer, P. Zarda, S. Mayer, and H. Weinfurter. The breakdown flash of silicon avalanche photodiodes–back door for eavesdropper attacks? *Journal of Modern Optics*, 48(13):2039–2047, 2001. doi: 10.1080/09500340110070235. URL http://www.informaworld.com/10.1080/09500340110070235.

[95] V. Zwiller, M. Pistol, D. Hessman, R. Cederström, W. Seifert, and L. Samuelson. Time-resolved studies of single semiconductor quantum dots. *Physical Review B (Condensed Matter and Materials Physics)*, 59(7):5021–5025, 1999. doi: 10.1103/physrevb.59.5021. URL http://link.aps.org/doi/10.1103/PhysRevB.59.5021.

[96] J. Gérard, B. Sermage, B. Gayral, B. Legrand, E. Costard, and V. Thierry-Mieg. Enhanced spontaneous emission by quantum boxes in a monolithic optical microcavity. *Physical Review Letters*, 81(5):1110–1113, 1998. doi: 10.1103/physrevlett.81.1110. URL http://link.aps.org/abstract/PRL/v81/p1110.

[97] G. S. Solomon, M. Pelton, and Y. Yamamoto. Single-mode spontaneous emission from a single quantum dot in a three-dimensional microcavity. *Physical Review Letters*, 86(17):3903–3906, 2001. doi: 10.1103/physrevlett.86.3903. URL http://link.aps.org/abstract/PRL/v86/p3903.

[98] A. Rastelli, A. Ulhaq, S. Kiravittaya, L. Wang, A. Zrenner, and O. G. Schmidt. In situ laser microprocessing of single self-assembled quantum dots and optical microcavities. *Applied Physics Letters*, 90(7):073120, 2007. doi: 10.1063/1.2431576. URL http://link.aip.org/link/?APL/90/073120/1.

[99] V. Zwiller and G. Björk. Improved light extraction from emitters in high refractive index materials using solid immersion lenses. *Journal of Applied Physics*, 92(2):660–665, 2002. doi: 10.1063/1.1487913. URL http://link.aip.org/link/?JAP/92/660/1.

[100] V. Zwiller, T. Aichele, W. Seifert, J. Persson, and O. Benson. Generating visible single photons on demand with single InP quantum dots. *Applied Physics Letters*, 82(10):1509–1511, 2003. doi: 10.1063/1.1558952. URL http://link.aip.org/link/?APL/82/1509/1.

[101] C. Santori, M. Pelton, G. S. Solomon, Y. Dale, and Y. Yamamoto. Triggered single photons from a quantum dot. *Physical Review Letters*, 86(8):1502–1505, 2001. doi: 10.1103/physrevlett.86.1502. URL http://link.aps.org/abstract/PRL/v86/p1502.

[102] J. Hecht. *Understanding Fibre Optics*. Prentice-Hall, London, 2002. ISBN 0-13-956145-5.

[103] G. N. Gol'tsman, O. Okunev, G. Chulkova, A. Lipatov, A. Semenov, K. Smirnov, B. Voronov, A. Dzardanov, C. P. Williams, and R. Sobolewski. Picosecond superconducting single-photon optical detector. *Applied Physics Letters*, 79(6):705–707, 2001. doi: 10.1063/1.1388868. URL http://link.aip.org/link/?APL/79/705/1.

[104] K. M. Rosfjord, J. K.W. Yang, E. A. Dauler, A. J. Kerman, V. Anant, B. Voronov, G. N. Gol'tsman, and K. K. Berggren. Nanowire single-photon detector with an integrated optical cavity and anti-reflection coating. *Optics Express*, 14(2):527–534, 2006. doi: 10.1364/opex.14.000527. URL http://www.opticsexpress.org/abstract.cfm?URI=oe-14-2-527.

[105] A. Divochiy, F. Marsili, D. Bitauld, A. Gaggero, R. Leoni, F. Mattioli, A. Korneev, V. Seleznev, N. Kaurova, O. Minaeva, G. N. Gol'tsman, K. G. Lagoudakis, M. Benkhaoul, F. Levy, and A. Fiore. Superconducting nanowire photon-number-resolving detector at telecommunication wavelengths. *Nature Photonics*, 2(5):302–306, 2008. doi: 10.1038/nphoton.2008.51. URL http://dx.doi.org/10.1038/nphoton.2008.51.

[106] J. Persson, M. Holm, C. Pryor, D. Hessman, W. Seifert, L. Samuelson, and M. Pistol. Optical and theoretical investigations of small InP quantum dots in $Ga_xIn_{1-x}P$. *Physical Review B (Condensed Matter and Materials Physics)*, 67(3):035320, January 2003. doi: 10.1103/physrevb.67.035320. URL http://link.aps.org/doi/10.1103/PhysRevB.67.035320.

[107] M. A. Nielsen and I. L. Chuang. *Quantum Computation and Quantum Information*. Cambridge University Press, 2000. ISBN 978-0521635035.

[108] I. L. Chuang, L. M. K. Vandersypen, X. Zhou, D. W. Leung, and S. Lloyd. Experimental realization of a quantum algorithm. *Nature*, 393(6681):143–146, 1998. doi: 10.1038/30181. URL http://dx.doi.org/10.1038/30181.

[109] J. A. Jones, M. Mosca, and R. H. Hansen. Implementation of a quantum search algorithm on a quantum computer. *Nature*, 393(6683): 344–346, 1998. doi: 10.1038/30687. URL http://dx.doi.org/10.1038/30687.

[110] P. Bianucci, A. Muller, C. K. Shih, Q. Q. Wang, Q. K. Xue, and C. Piermarocchi. Experimental realization of the one qubit Deutsch-Jozsa algorithm in a quantum dot. *Physical Review B (Condensed Matter and Materials Physics)*, 69(16):161303, 2004. doi: 10.1103/physrevb.69.161303. URL http://link.aps.org/abstract/PRB/v69/e161303.

[111] S. Gulde, M. Riebe, G. P. T. Lancaster, C. Becher, J. Eschner, H. Häffner, F. Schmidt-Kaler, I. L. Chuang, and R. Blatt. Implementation of the Deutsch-Jozsa algorithm on an ion-trap quantum computer. *Nature*, 421(6918):48–50, 2003. doi: 10.1038/nature01336. URL http://dx.doi.org/10.1038/nature01336.

[112] S. Takeuchi. Experimental demonstration of a three-qubit quantum computation algorithm using a single photon and linear optics. *Physical Review A (Atomic, Molecular, and Optical Physics)*, 62(3):032301, 2000. doi: 10.1103/physreva.62.032301. URL http://link.aps.org/doi/10.1103/PhysRevA.62.032301.

[113] E. Brainis, L.-P. Lamoureux, N. J. Cerf, P. Emplit, M. Haelterman, and S. Massar. Fiber-optics implementation of the Deutsch-Jozsa and Bernstein-Vazirani quantum algorithms with three qubits. *Physical Review Letters*, 90(15):157902, 2003. doi: 10.1103/physrevlett.90.157902. URL http://link.aps.org/doi/10.1103/PhysRevLett.90.157902.

[114] M. Mohseni, J. S. Lundeen, K. J. Resch, and A. M. Steinberg. Experimental application of decoherence-free subspaces in an optical quantum-computing algorithm. *Physical Review Letters*, 91(18): 187903, 2003. doi: 10.1103/physrevlett.91.187903. URL http://link.aps.org/doi/10.1103/PhysRevLett.91.187903.

[115] M. Scholz, T. Aichele, S. Ramelow, and O. Benson. Deutsch-Jozsa algorithm using triggered single photons from a single quantum dot. *Physical Review Letters*, 96(18):180501, 2006. doi: 10.1103/physrevlett.96.180501. URL http://link.aps.org/abstract/PRL/v96/e180501.

[116] V. Zwiller, T. Aichele, and O. Benson. Single-photon Fourier spectroscopy of excitons and biexcitons in single quantum dots. *Physical Review B (Condensed Matter and Materials Physics)*, 69(16):165307, 2004. doi: 10.1103/physrevb.69.165307. URL http://link.aps.org/abstract/PRB/v69/e165307.

[117] W. H. Zurek, E. Knill, and R. Laflamme. Resilient quantum computation: Error models and thresholds. *Proceedings of the Royal Society A: Mathematical, Physical and Engineering Sciences*, 454(1969):365–384, 1998. doi: 10.1098/rspa.1998.0166. URL http://dx.doi.org/10.1098/rspa.1998.0166.

[118] B. M. Terhal and G. Burkard. Fault-tolerant quantum computation for local non-Markovian noise. *Physical Review A (Atomic, Molecular, and Optical Physics)*, 71(1):012336, 2005. doi: 10.1103/physreva.71.012336. URL http://link.aps.org/abstract/PRA/v71/e012336.

[119] P. Zanardi and M. Rasetti. Noiseless quantum codes. *Physical Review Letters*, 79(17):3306–3309, 1997. doi: 10.1103/physrevlett.79.3306. URL http://link.aps.org/doi/10.1103/PhysRevLett.79.3306.

[120] D. A. Lidar, I. L. Chuang, and K. B. Whaley. Decoherence-free subspaces for quantum computation. *Physical Review Letters*, 81 (12):2594–2597, 1998. doi: 10.1103/physrevlett.81.2594. URL http://link.aps.org/doi/10.1103/PhysRevLett.81.2594.

[121] R. Lettow, V. Ahtee, R. Pfab, A. Renn, E. Ikonen, S. Götzinger, and V. Sandoghdar. Realization of two Fourier-limited solid-state single-photon sources. *Optics Express*, 15(24):15842–15847, 2007. doi: 10.1364/oe.15.015842. URL http://www.opticsexpress.org/abstract.cfm?URI=oe-15-24-15842.

[122] O. Benson, C. Santori, M. Pelton, and Y. Yamamoto. Regulated and entangled photons from a single quantum dot. *Physical Review Letters*, 84(11):2513–2516, 2000. doi: 10.1103/physrevlett.84.2513. URL http://link.aps.org/abstract/PRL/v84/p2513.

[123] R. M. Stevenson, R. J. Young, P. Atkinson, K. Cooper, D. A. Ritchie, and A. J. Shields. A semiconductor source of triggered entangled photon pairs. *Nature*, 439(7073):179–182, 2006. doi: 10.1038/nature04446. URL http://dx.doi.org/10.1038/nature04446.

[124] N. Akopian, N. H. Lindner, E. Poem, Y. Berlatzky, J. Avron, D. Gershoni, B. D. Gerardot, and P. M. Petroff. Entangled photon pairs from

semiconductor quantum dots. *Physical Review Letters*, 96(13):130501, 2006. doi: 10.1103/physrevlett.96.130501. URL http://link.aps.org/abstract/PRL/v96/e130501.

[125] H. de Riedmatten, I. Marcikic, J. A. W. van Houwelingen, W. Tittel, H. Zbinden, and N. Gisin. Long-distance entanglement swapping with photons from separated sources. *Physical Review A (Atomic, Molecular, and Optical Physics)*, 71(5):050302, 2005. doi: 10.1103/physreva.71.050302. URL http://link.aps.org/abstract/PRA/v71/e050302.

[126] A. Poppe, A. Fedrizzi, R. Ursin, H. Böhm, T. Lörunser, O. Maurhardt, M. Peev, M. Suda, C. Kurtsiefer, H. Weinfurter, T. Jennewein, and A. Zeilinger. Practical quantum key distribution with polarization entangled photons. *Optics Express*, 12(16):3865–3871, 2004. doi: 10.1364/opex.12.003865. URL http://www.opticsexpress.org/abstract.cfm?URI=oe-12-16-3865.

[127] J. L. O'Brien, G. J. Pryde, A. G. White, T. C. Ralph, and D. Branning. Demonstration of an all-optical quantum controlled-not gate. *Nature*, 426(6964):264–267, 2003. doi: 10.1038/nature02054. URL http://dx.doi.org/10.1038/nature02054.

[128] T. Kimura, Y. Nambu, T. Hatanaka, A. Tomita, H. Kosaka, and K. Nakamura. Single-photon interference over 150 km transmission using silica-based integrated-optic interferometers for quantum cryptography. *Japanese Journal of Applied Physics*, 43(9):L1217–L1219, 2004. doi: 10.1143/jjap.43.l1217. URL http://jjap.ipap.jp/link?JJAP/43/L1217/.

[129] J. B. Kim, O. Benson, H. Kan, and Y. Yamamoto. A single-photon turnstile device. *Nature*, 397(6719):500–503, 1999. doi: 10.1038/17295. URL http://dx.doi.org/10.1038/17295.

[130] A. Fiore, J. X. Chen, and M. Ilegems. Scaling quantum-dot light-emitting diodes to submicrometer sizes. *Applied Physics Letters*, 81(10):1756–1758, 2002. doi: 10.1063/1.1504880. URL http://link.aip.org/link/?APL/81/1756/1.

[131] C. Zinoni, B. Alloing, C. Paranthoën, and A. Fiore. Three-dimensional wavelength-scale confinement in quantum dot microcavity light-emitting diodes. *Applied Physics Letters*, 85(12):2178–2180, 2004. doi: 10.1063/1.1791341. URL http://link.aip.org/link/?APL/85/2178/1.

[132] D. J. P. Ellis, A. J. Bennett, A. J. Shields, P. Atkinson, and D. A. Ritchie. Electrically addressing a single self-assembled quantum dot. *Applied Physics Letters*, 88(13):133509, 2006. doi: 10.1063/1.2190451. URL http://link.aip.org/link/?APL/88/133509/1.

[133] R. Schmidt, U. Scholz, M. Vitzethum, R. Fix, C. Metzner, P. Kailuweit, D. Reuter, A. Wieck, M. C. Hübner, S. Stufler, A. Zrenner, S. Malzer, and G. H. Döhler. Fabrication of genuine single-quantum-dot light-emitting diodes. *Applied Physics Letters*, 88(12):121115, 2006. doi: 10.1063/1.2188057. URL http://link.aip.org/link/?APL/88/121115/1.

[134] C. Monat, B. Alloing, C. Zinoni, L. H. Li, and A. Fiore. Nanostructured current-confined single quantum dot light-emitting diode at 1300 nm. *Nano Letters*, 6(7):1464–1467, 2006. doi: 10.1021/nl060800t. URL http://pubs.acs.org/doi/abs/10.1021/nl060800t.

[135] A. Lochmann, E. Stock, O. Schulz, F. Hopfer, D. Bimberg, V. A. Haisler, A. I. Toropov, A. K. Bakarov, M. Scholz, S. Büttner, and O. Benson. Electrically driven quantum dot single photon source. *Physica Status Solidi (c)*, 4(2):547–550, 2007. doi: 10.1002/pssc.200673201. URL http://dx.doi.org/10.1002/pssc.200673201.

[136] A. Lochmann, E. Stock, O. Schulz, F. Hopfer, D. Bimberg, V. A. Haisler, A. I. Toropov, A. K. Bakarov, and A. K. Kalagin. Electrically driven single quantum dot polarised single photon emitter. *Electronics Letters*, 42(13):774–775, 2006. doi: 10.1049/el:20061076. URL http://ieeexplore.ieee.org/stamp/stamp.jsp?tp=&isnumber=34564&arnumber=1648582.

[137] V. A. Haisler, F. Hopfer, R. L. Sellin, A. Lochmann, K. Fleischer, N. Esser, W. Richter, N. N. Ledentsov, D. Bimberg, C. Möller, and N. Grote. Micro-raman studies of vertical-cavity surface-emitting lasers with Al_xO_y/GaAs distributed bragg reflectors. *Applied Physics Letters*, 81(14):2544–2546, 2002. doi: 10.1063/1.1511533. URL http://link.aip.org/link/?APL/81/2544/1.

[138] P. Michler. *Single Quantum Dots*. Springer, Berlin Heidelberg, 2003. ISBN 978-3-540-14022-1.

[139] R. Heitz, A. Kalburge, Q. Xie, M. Grundmann, P. Chen, A. Hoffmann, A. Madhukar, and D. Bimberg. Excited states and energy relaxation in stacked InAs/GaAs quantum dots. *Physical Review B*

(Condensed Matter and Materials Physics), 57(15):9050–9060, 1998. doi: 10.1103/physrevb.57.9050. URL `http://link.aps.org/doi/10.1103/PhysRevB.57.9050`.

[140] R. M. Thompson, R. M. Stevenson, A. J. Shields, I. Farrer, C. J. Lobo, D. A. Ritchie, M. L. Leadbeater, and M. Pepper. Single-photon emission from exciton complexes in individual quantum dots. *Physical Review B (Condensed Matter and Materials Physics)*, 64(20):201302, 2001. doi: 10.1103/physrevb.64.201302. URL `http://link.aps.org/doi/10.1103/PhysRevB.64.201302`.

[141] C. Becher, A. Kiraz, P. Michler, W. V. Schönfeld, P. M. Petroff, L. Zhang, E. Hu, and A. Imamoglu. A quantum dot single-photon source. *Physica E: Low-dimensional Systems and Nanostructures*, 13(2-4):412–417, 2002. doi: 10.1016/s1386-9477(02)00156-x. URL `http://www.sciencedirect.com/science/article/B6VMT-44Y0YTW-W/2/f605511421eb6a8f90d9e059f4463750`.

[142] R. Seguin, A. Schliwa, S. Rodt, K. Pötschke, U. W. Pohl, and D. Bimberg. Size-dependent fine-structure splitting in self-organized InAs/GaAs quantum dots. *Physical Review Letters*, 95(25):257402, 2005. doi: 10.1103/physrevlett.95.257402. URL `http://link.aps.org/abstract/PRL/v95/e257402`.

[143] R. W. Boyd. *Nonlinear Optics*. Academic Press, San Diego, 2002. ISBN 978-0121216825.

[144] V. G. Dmitriev, G. G. Gurzadyan, and D. N. Nikogosyan. *Handbook of Nonlinear Optical Crystals*. Springer, Berlin, 1999. ISBN 978-3540653943.

[145] F. Brehat and B. Wyncke. Calculation of double-refraction walk-off angle along the phase-matching directions in non-linear biaxial crystals. *Journal of Physics B: Atomic, Molecular and Optical Physics*, 22(11):1891–1898, 1989. doi: 10.1088/0953-4075/22/11/020. URL `http://stacks.iop.org/0953-4075/22/1891`.

[146] M. Houe and P. D. Townsend. An introduction to methods of periodic poling for second-harmonic generation. *Journal of Physics D: Applied Physics*, 28(9):1747–1763, 1995. doi: 10.1088/0022-3727/28/9/001. URL `http://stacks.iop.org/0022-3727/28/1747`.

[147] D. A. Kleinman, A. Ashkin, and G. D. Boyd. Second-harmonic generation of light by focused laser beams. *Physical Review*, 145(1):338–379,

1966. doi: 10.1103/physrev.145.338. URL http://link.aps.org/doi/10.1103/PhysRev.145.338.

[148] G. D. Boyd, A. Ashkin, J. M. Dziedzic, and D. A. Kleinman. Second-harmonic generation of light with double refraction. *Physical Review*, 137(4A):A1305–A1320, 1965. doi: 10.1103/physrev.137.a1305. URL http://link.aps.org/doi/10.1103/PhysRev.137.A1305.

[149] G. D. Boyd and D. A. Kleinman. Parametric interaction of focused Gaussian light beams. *Journal of Applied Physics*, 39(8):3597–3639, 1968. doi: 10.1063/1.1656831. URL http://link.aip.org/link/?JAP/39/3597/1.

[150] B. E. A. Saleh and M. C. Teich. *Grundlagen der Photonik*. Wiley-VCH, Weinheim, 2008. ISBN 9783527406777.

[151] G. A. Reider. *Photonik - Eine Einführung in die Grundlagen*. Springer-Verlag, Berlin, Heidelberg, New York, 2004. ISBN 978-3211219010.

[152] R. V. Pound. Electronic frequency stabilization of microwave oscillators. *Review of Scientific Instruments*, 17(11):490–505, 1946. doi: 10.1063/1.1770414. URL http://link.aip.org/link/?RSI/17/490/1.

[153] R. W. P. Drever, J. L. Hall, F. V. Kowalski, J. Hough, G. M. Ford, A. J. Munley, and H. Ward. Laser phase and frequency stabilization using an optical resonator. *Applied Physics B: Lasers and Optics*, 31(2):97–105, 1983. doi: 10.1007/bf00702605. URL http://dx.doi.org/10.1007/BF00702605.

[154] E. D. Black. An introduction to Pound-Drever-Hall laser frequency stabilization. *American Journal of Physics*, 69(1):79–87, 2001. doi: 10.1119/1.1286663. URL http://dx.doi.org/10.1119/1.1286663.

[155] T. Hänsch and B. Couillaud. Laser frequency stabilization by polarization spectroscopy of a reflecting reference cavity. *Optics Communications*, 35(3):441–444, 1980. doi: 10.1016/0030-4018(80)90069-3. URL http://www.sciencedirect.com/science/article/B6TVF-46FR1KK-TC/2/18877ced11006bcbf37db8288bcafd7f.

[156] R. C. Jones. New calculus for the treatment of optical systems. VIII. Electromagnetic theory. *Journal of the Optical Society of America*, 46(2):126–131, 1956. doi: 10.1364/josa.46.000126. URL http://www.opticsinfobase.org/abstract.cfm?URI=josa-46-2-126.

[157] W. Demtröder. *Laser Spectroscopy. Basic Concepts and Instrumenta-tion.* Spinger, Berlin, 2007. ISBN 978-3540337928.

[158] G. C. Bjorklund. Frequency-modulation spectroscopy: A new method for measuring weak absorptions and dispersions. *Optics Letters*, 5(1): 15–17, 1980. doi: 10.1364/ol.5.000015. URL http://ol.osa.org/abstract.cfm?URI=ol-5-1-15.

[159] G. C. Bjorklund, M. D. Levenson, W. Lenth, and C. Ortiz. Frequency modulation (FM) spectroscopy. *Applied Physics B: Lasers and Optics*, 32(3):145–152, 1983. doi: 10.1007/bf00688820. URL http://dx.doi.org/10.1007/BF00688820.

[160] O. S. Brozek. Effiziente Frequenzverdopplung mit Diodenlasern. Mas-ter's thesis, Universität Hannover, 1995.

[161] H. Mabuchi, E. S. Polzik, and H. J. Kimble. Blue-light induced infrared absorption in $KNbO_3$. *Journal of the Optical Society of America B*, 11(10):2023–2029, 1994. doi: 10.1364/josab.11.002023. URL http://josab.osa.org/abstract.cfm?URI=josab-11-10-2023.

[162] A. E. B. Nielsen and K. Mølmer. Single-photon-state generation from a continuous-wave nondegenerate optical parametric oscillator. *Physi-cal Review A (Atomic, Molecular, and Optical Physics)*, 75(2):023806, 2007. doi: 10.1103/physreva.75.023806. URL http://link.aps.org/abstract/PRA/v75/e023806.

[163] A. E. B. Nielsen and K. Mølmer. Multimode analysis of the light emitted from a pulsed optical parametric oscillator. *Physical Review A (Atomic, Molecular, and Optical Physics)*, 76(3):033832, 2007. doi: 10.1103/physreva.76.033832. URL http://link.aps.org/abstract/PRA/v76/e033832.

[164] K. J. Blow, R. Loudon, S. J. D. Phoenix, and T. J. Shepherd. Contin-uum fields in quantum optics. *Physical Review A (Atomic, Molecular, and Optical Physics)*, 42(7):4102–4114, 1990. doi: 10.1103/physreva.42.4102. URL http://link.aps.org/doi/10.1103/PhysRevA.42.4102.

[165] U. Herzog, M. Scholz, and O. Benson. Theory of biphoton generation in a single-resonant optical parametric oscillator far below threshold. *Physical Review A (Atomic, Molecular, and Optical Physics)*, 77(2): 023826, 2008. doi: 10.1103/physreva.77.023826. URL http://link.aps.org/abstract/PRA/v77/e023826.

[166] M. J. Collett and C. W. Gardiner. Squeezing of intracavity and traveling-wave light fields produced in parametric amplification. *Physical Review A (Atomic, Molecular, and Optical Physics)*, 30(3):1386–1391, 1984. doi: 10.1103/physreva.30.1386. URL http://link.aps.org/abstract/PRA/v30/p1386.

[167] C. K. Hong and L. Mandel. Theory of parametric frequency down conversion of light. *Physical Review A (Atomic, Molecular, and Optical Physics)*, 31(4):2409–2418, 1985. doi: 10.1103/physreva.31.2409. URL http://link.aps.org/abstract/PRA/v31/p2409.

[168] R. Ghosh, C. K. Hong, Z. Y. Ou, and L. Mandel. Interference of two photons in parametric down conversion. *Physical Review A (Atomic, Molecular, and Optical Physics)*, 34(5):3962–3968, 1986. doi: 10.1103/physreva.34.3962. URL http://link.aps.org/doi/10.1103/PhysRevA.34.3962.

[169] H. Hellwig, J. Liebertz, and L. Bohaty. Linear optical properties of the monoclinic bismuth borate BiB_3O_6. *Journal of Applied Physics*, 88(1):240–244, 2000. doi: 10.1063/1.373647. URL http://link.aip.org/link/?JAP/88/240/1.

[170] M. Ghotbi and M. Ebrahim-Zadeh. Optical second harmonic generation properties of BiB_3O_6. *Optics Express*, 12(24):6002–6019, 2004. doi: 10.1364/opex.12.006002. URL http://www.opticsexpress.org/abstract.cfm?URI=oe-12-24-6002.

[171] M. Oberparleiter. *Effiziente Erzeugung verschränkter Photonenpaare*. PhD thesis, Ludwig-Maximilians-Universität München, 2002.

[172] Y. J. Lu, R. L. Campbell, and Z. Y. Ou. Mode-locked two-photon states. *Physical Review Letters*, 91(16):163602, 2003. doi: 10.1103/physrevlett.91.163602. URL http://link.aps.org/abstract/PRL/v91/e163602.

[173] T. Y. Fan, C. E. Huang, B. Q. Hu, R. C. Eckardt, Y. X. Fan, R. L. Byer, and R. S. Feigelson. Second harmonic generation and accurate index of refraction measurements in flux-grown $KTiOPO_4$. *Applied Optics*, 26(12):2390–2394, 1987. doi: 10.1364/ao.26.002390. URL http://ao.osa.org/abstract.cfm?URI=ao-26-12-2390.

[174] K. Kato and E. Takaoka. Sellmeier and thermo-optic dispersion formulas for KTP. *Applied Optics*, 41(24):5040–5044, 2002. doi:

10.1364/ao.41.005040. URL http://ao.osa.org/abstract.cfm?URI=ao-41-24-5040.

[175] P. D. Drummond and M. D. Reid. Correlations in nondegenerate parametric oscillation. II. Below threshold results. *Physical Review A (Atomic, Molecular, and Optical Physics)*, 41(7):3930–3949, 1990. doi: 10.1103/physreva.41.3930. URL http://link.aps.org/doi/10.1103/PhysRevA.41.3930.

[176] K. J. McNeil and C. W. Gardiner. Quantum statistics of parametric oscillation. *Physical Review A (Atomic, Molecular, and Optical Physics)*, 28(3):1560–1566, 1983. doi: 10.1103/physreva.28.1560. URL http://link.aps.org/doi/10.1103/PhysRevA.28.1560.

[177] S. Fasel, O. Alibart, S. Tanzilli, P. Baldi, A. Beveratos, N. Gisin, and H. Zbinden. High-quality asynchronous heralded single-photon source at telecom wavelength. *New Journal of Physics*, 6:163–173, 2004. doi: 10.1088/1367-2630/6/1/163. URL http://stacks.iop.org/1367-2630/6/163.

[178] E. Bocquillon, C. Couteau, M. Razavi, R. Laflamme, and G. Weihs. Coherence measures for heralded single-photon sources. *e-print arXiv quant-ph*, 2008. URL http://arxiv.org/abs/0807.1725.

[179] F. N. C. Wong, J. H. Shapiro, and T. Kim. Efficient generation of polarization-entangled photons in a nonlinear crystal. *Laser Physics*, 16(11):1517–1524, 2006. doi: 10.1134/s1054660x06110053. URL http://dx.doi.org/10.1134/S1054660X06110053.

[180] J. L. Hall, M. S. Taubman, and J. Ye. Laser stabilization. Technical report, JILA, 1999.

[181] T. Toyoda and M. Yabe. The temperature dependence of the refractive indices of fused silica and crystal quartz. *Journal of Physics D: Applied Physics*, 16(5):L97–L100, 1983. doi: 10.1088/0022-3727/16/5/002. URL http://stacks.iop.org/0022-3727/16/L97.

[182] C. H. Bennett, F. Bessette, G. Brassard, L. Salvail, and J. Smolin. Experimental quantum cryptography. *Journal of Cryptology*, 5(1):3–28, 1992. doi: 10.1007/bf00191318. URL http://dx.doi.org/10.1007/BF00191318.

[183] T. Aichele, G. Reinaudi, and O. Benson. Separating cascaded photons from a single quantum dot: Demonstration of multiplexed quantum

cryptography. *Physical Review B (Condensed Matter and Materials Physics)*, 70(23):235329, 2004. doi: 10.1103/physrevb.70.235329. URL http://link.aps.org/doi/10.1103/PhysRevB.70.235329.

[184] S. Rebic, D. Vitali, C. Ottaviani, P. Tombesi, M. Artoni, F. S. Cataliotti, and R. Corbalán. Polarization phase gate with a tripod atomic system. *Physical Review A (Atomic, Molecular, and Optical Physics)*, 70(3):032317, 2004. doi: 10.1103/physreva.70.032317. URL http://link.aps.org/abstract/PRA/v70/e032317.

[185] C. Ottaviani, D. Vitali, M. Artoni, F. S. Cataliotti, and P. Tombesi. Polarization qubit phase gate in driven atomic media. *Physical Review Letters*, 90(19):197902, 2003. doi: 10.1103/physrevlett.90.197902. URL http://link.aps.org/abstract/PRL/v90/e197902.

[186] G. Alzetta, A. Gozzini, L. Moi, and G. Orriols. An experimental method for the observation of the RF transitions in oriented Na vapor. *Nuovo Cimento B*, 36(1):5–20, 1976. doi: 10.1007/bf02749417. URL http://dx.doi.org/10.1007/BF02749417.

[187] R. M. Whitley and C. R. Stroud, Jr. Double optical resonance. *Physical Review A (Atomic, Molecular, and Optical Physics)*, 14(4):1498–1513, 1976. doi: 10.1103/physreva.14.1498. URL http://link.aps.org/doi/10.1103/PhysRevA.14.1498.

[188] H. R. Gray, R. M. Whitley, and C. R. Stroud, Jr. Coherent trapping of atomic populations. *Optics Letters*, 3(6):218–220, 1978. doi: 10.1364/ol.3.000218. URL http://ol.osa.org/abstract.cfm?URI=ol-3-6-218.

[189] S. E. Harris, J. E. Field, and A. Imamoglu. Nonlinear optical processes using electromagnetically induced transparency. *Physical Review Letters*, 64(10):1107–1110, 1990. doi: 10.1103/physrevlett.64.1107. URL http://link.aps.org/abstract/PRL/v64/p1107.

[190] A. Kasapi, M. Jain, G. Y. Yin, and S. E. Harris. Electromagnetically induced transparency: Propagation dynamics. *Physical Review Letters*, 74(13):2447–2450, 1995. doi: 10.1103/physrevlett.74.2447. URL http://link.aps.org/abstract/PRL/v74/p2447.

[191] S. E. Harris. Electromagnetically induced transparency. *Physics Today*, 50(7):36–42, 1997. doi: 10.1063/1.881806. URL http://link.aip.org/link/?PTO/50/36/1.

[192] A. S. Zibrov, M. D. Lukin, D. E. Nikonov, L. Hollberg, M. O. Scully, V. L. Velichansky, and H. G. Robinson. Experimental demonstration of laser oscillation without population inversion via quantum interference in Rb. *Physical Review Letters*, 75(8):1499–1502, 1995. doi: 10.1103/physrevlett.75.1499. URL http://link.aps.org/abstract/PRL/v75/p1499.

[193] G. Grynberg, M. Pinard, and P. Mandel. Amplification without population inversion in a V three-level system: A physical interpretation. *Physical Review A (Atomic, Molecular, and Optical Physics)*, 54(1):776–785, 1996. doi: 10.1103/physreva.54.776. URL http://link.aps.org/abstract/PRA/v54/p776.

[194] C. Fort, F. S. Cataliotti, T. Hänsch, M. Inguscio, and M. Prevedelli. Gain without inversion on the cesium D1 line. *Optics Communications*, 139(1):31–34, 1997. doi: 10.1016/s0030-4018(97)00094-1. URL http://www.sciencedirect.com/science/article/B6TVF-3W0K0VV-12/2/9cfa9934d1192e31f1b9c9fc2a34b578.

[195] O. Kocharovskaya, Y. V. Rostovtsev, and M. O. Scully. Stopping light via hot atoms. *Physical Review Letters*, 86(4):628–631, 2001. doi: 10.1103/physrevlett.86.628. URL http://link.aps.org/abstract/PRL/v86/p628.

[196] D. F. Phillips, A. Fleischhauer, A. Mair, R. L. Walsworth, and M. D. Lukin. Storage of light in atomic vapor. *Physical Review Letters*, 86(5):783–786, 2001. doi: 10.1103/physrevlett.86.783. URL http://link.aps.org/abstract/PRL/v86/p783.

[197] C. Cohen-Tannoudji, J. Dupont-Roc, and G. Grynberg. *Atom-Photon Interactions: Basic Processes and Applications*. Wiley VCH, New York, 1998. ISBN 978-0471293361.

[198] H. Lee, Y. V. Rostovtsev, C. J. Bednar, and A. Javan. From laser-induced line narrowing to electro magnetically induced transparency: Closed system analysis. *Applied Physics B: Lasers and Optics*, 76(1): 33–39, 2003. doi: 10.1007/s00340-002-1030-5. URL http://dx.doi.org/10.1007/s00340-002-1030-5.

[199] E. Figueroa, F. Vewinger, J. Appel, and A. I. Lvovsky. Decoherence of electromagnetically induced transparency in atomic vapor. *Optics Letters*, 31(17):2625–2627, 2006. doi: 10.1364/ol.31.002625. URL http://www.opticsinfobase.org/abstract.cfm?URI=ol-31-17-2625.

[200] B. Steinheil. Aufbau eines frequenzverdoppelten Diodenlasersystems zur Untersuchung ultrakalter Chrom-Atome. Master's thesis, Universität Stuttgart, 2003.

[201] A. Javan, O. Kocharovskaya, H. Lee, and M. O. Scully. Narrowing of electromagnetically induced transparency resonance in a Doppler-broadened medium. *Physical Review A (Atomic, Molecular, and Optical Physics)*, 66(1):013805, 2002. doi: 10.1103/physreva.66.013805. URL http://link.aps.org/abstract/PRA/v66/e013805.

[202] L. Cacciapuoti, M. de Angelis, M. Fattori, G. Lamporesi, T. Petelski, M. Prevedelli, J. Stuhler, and G. Tino. Analog digital phase and frequency detector for phase locking of diode lasers. *Review of Scientific Instruments*, 76(5):053111, 2005. doi: 10.1063/1.1914785. URL http://link.aip.org/link/?RSINAK/76/053111/1.

[203] M. Prevedelli. Analog digital phase and frequency detector. Technical report, University of Bologna, 2002. private communication.

[204] T. E. Sterne. Multi-lamellar cylindrical magnetic shields. *Review of Scientific Instruments*, 6(10):324–326, 1935. doi: 10.1063/1.1751884. URL http://link.aip.org/link/?RSINAK/6/324/1.

[205] D. Höckel, M. Scholz, and O. Benson. A robust phase-locked diode laser system for EIT experiments in cesium. *Applied Physics B: Lasers and Optics*, 94(3):429, 2009. doi: 10.1007/s00340-008-3313-y. URL http://www.springerlink.com/content/g32k308148892701/fulltext.pdf.

[206] A. V. Turukhin, V. S. Sudarshanam, M. S. Shahriar, J. A. Musser, B. S. Ham, and P. R. Hemmer. Observation of ultraslow and stored light pulses in a solid. *Physical Review Letters*, 88(2):023602, 2001. doi: 10.1103/physrevlett.88.023602. URL http://link.aps.org/abstract/PRL/v88/e023602.

[207] P. Kolchin, C. Belthangady, S. Du, G. Y. Yin, and S. E. Harris. Electro-optic modulation of single photons. *Physical Review Letters*, 101(10):103601, 2008. doi: 10.1103/physrevlett.101.103601. URL http://link.aps.org/abstract/PRL/v101/e103601.

[208] M. D. Eisaman, L. Childress, A. André, F. Massou, A. S. Zibrov, and M. D. Lukin. Shaping quantum pulses of light via coherent atomic memory. *Physical Review Letters*, 93(23):233602, 2004. doi: 10.1103/

physrevlett.93.233602. URL http://link.aps.org/abstract/PRL/v93/e233602.

[209] J. K. Thompson, J. Simon, H. Loh, and V. Vuletic. A high-brightness source of narrowband, identical-photon pairs. *Science*, 313(5783):74–77, 2006. doi: 10.1126/science.1127676. URL http://www.sciencemag.org/cgi/content/abstract/313/5783/74.

[210] D. A. Steck. Cesium D Line Data. Technical report, Los Alamos National Laboratory, 1998.

Appendix A

Cesium D1 Line

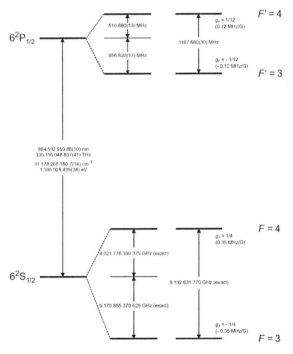

Figure A.1: Frequency splittings between the hyperfine energy levels on the cesium D1 line. The ground-state values are exact, as a result of the current definition of the second. The g_F denote the approximate Landé factors for each level, together with level shifts due to a magnetic field [210].

167

Appendix B

Views of the Lab

Figure B.1: Photograph of the EIT cell with opened three-layer magnetic shielding.

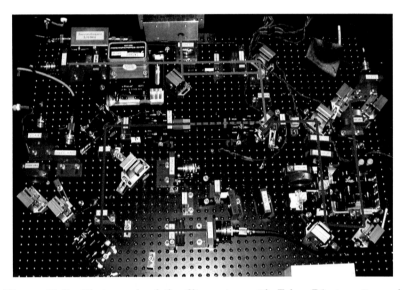

Figure B.2: Photograph of the filter setup with Fabry-Pérot cavity and etalon. The main beam paths of DRO signal photons (red) and reference laser (orange) are sketched.

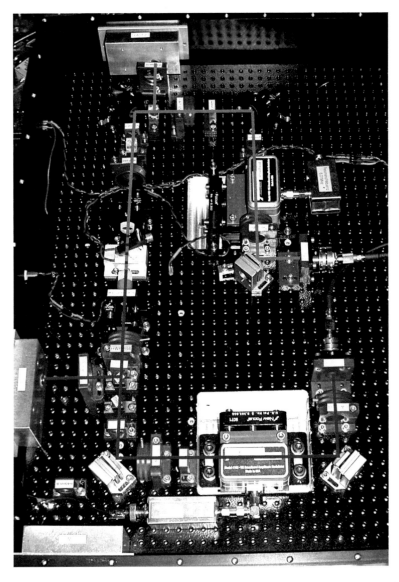

Figure B.3: Photograph of the transfer setup. Again, master laser (red) and reference (orange) beams are depicted.

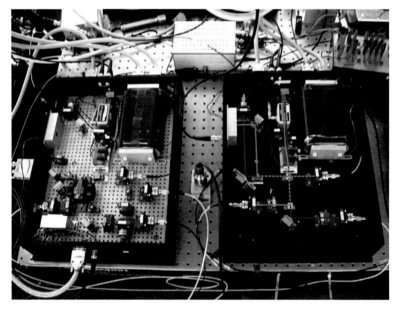

Figure B.4: Photograph of the time-bin interferometers and fiber links. DRO signal photons (red) and reference laser (orange) are split inside a calcite beam displacer.

Abbreviations

APD	Avalanche Photo Diode
BiBO	Bismuth Triborate
BP	Band-Pass
BS	Beam Splitter
BW	Bandwidth
CPLD	Complex Programmable Logic Device
cw	continuous-wave
DL	Diode Laser
DM	Dichroic Mirror
DRO	Double-Resonant OPO
EIT	Electromagnetically Induced Transparency
EOM	Electro-Optic Modulator
FC	Fiber Coupler
FFT	Fast Fourier Transform
FMS	Frequency Modulation Spectroscopy
FSR	Free Spectral Range
FWHM	Full Width at Half Maximum
HBT	Hanbury-Brown and Twiss
HWP	Half-Wave Plate
IF	Interference Filter
$KNbO_3$	Potassium Niobate
KTP	Potassium Titanyl Phosphate
LOQC	Linear Optics Quantum Computation
KLM	Knill, Laflamme, and Milburn
LED	Light Emitting Diode
LO	Local Oscillator
LP	Long-Pass
MBE	Molecular Beam Epitaxy
MOCVD	Metal-Organic Chemical Vapor Deposition
NA	Numerical Aperture
OPO	Optical Parametric Oscillator

PBS	Polarizing Beam Splitter
PD	Photo Diode
PDH	Pound-Drever-Hall
PH	Pinhole
PMF	Polarization-Maintaining Fiber
Pol	Polarizer
PPKTA	Periodically Poled Potassium Titanyl Arsenate
PPKTP	Periodically Poled Potassium Titanyl Phosphate
PPLN	Periodically Poled Lithium Niobate
PPLT	Periodically Poled Lithium Tantalate
PLL	Phase-Locked Loop
PS	Power Source
PZT	Piezo-electric Tube
QD	Quantum Dot
QIP	Quantum Information Processing
QWP	Quarter-Wave Plate
RF	Radio Frequency
SHG	Second Harmonic Generation
SPDC	Spontaneous Parametric Down-Conversion
SPS	Single-Photon Source
SRO	Single-Resonant OPO
TEM	Transverse Electro-Magnetic
TFP	Thin-Film Polarizer
X	Exciton
XX	Biexciton

Acknowledgements

First of all I would like to express my sincere gratitude to Professor Oliver Benson who has been my supervisor since the beginning of my graduate studies. He provided me with many helpful suggestions, important advice, and constant encouragement during the course of this thesis. I thank him for his continuous trust in my skills which gave me extensive freedom during my work.

Dr. Thomas Aichele has been my friend for many years. I am happy he returned to the path of academic research after his one-year excursion into the real world. During my first year, I learned to esteem his calm and steady approach and way of working, and we look back on a great time in Berlin. We will surely sail on a close reach together, again.

Major parts of this thesis would not have been possible without the persistent help and commitment of Lars Koch. He exceeded all expectations one may have of a diploma student, and I will miss our interesting discussions and joyful time in the lab and during coffee breaks.

Joining my project during her Bachelor thesis, Silvia Arroyo Camejo has become a close friend of mine. I admire her energetic devotion to settle only for the complete answer and relish our rambling conversations about physics, philosophy, and beyond.

For the solution of theoretical issues, Dr. Ulrike Herzog was an amazing help and source of ideas. I very much enjoy her company and look up to her wealth of experience.

Several diploma students worked on the diverse projects that are covered by this thesis. I especially want to thank Roland Ullmann, David Höckel, Florian Wolfgramm, Nils Neubauer, and Sven Büttner.

Klaus Palis lent me an ear whenever electronic challenges appeared. His contributions were crucial for the success of this work. I am also grateful to Alexander Walter who assembled numerous electronic devices, spending a night shift if requests seemed urgent.

I also thank all other members of the NANO, QOM, and AMO groups for the inspiring time at Hausvogteiplatz.

Evangelisches Studienwerk Villigst and Deutsche Telekom Stiftung sponsored my thesis, and I appreciate their financial aid and conceptual support.

Throughout all my studies, my father Dieter gave me his unconditional support for which I am very much indebted to him. Sina joined my life just half a year ago and has become the center of my life. She continuously encourages me to do my best, and I thank her for her love and patience.

Selbständigkeitserklärung

Hiermit erkläre ich, die vorliegende Arbeit selbständig ohne fremde Hilfe verfasst und nur die angegebene Literatur und Hilfsmittel verwendet zu haben. Ich habe mich anderwärts nicht um einen Doktorgrad beworben und besitze einen entsprechenden Doktorgrad nicht.
Ich erkläre die Kenntnisnahme der dem Verfahren zugrunde liegenden Promotionsordnung der Mathematisch-Naturwissenschaftlichen Fakultät I der Humboldt-Universität zu Berlin.

Berlin, den 22.01.2009

Matthias Scholz